# Characterization Techniques for Nanomaterials

Manipulation of matter at the nanoscale level is the key factor in nanotechnology, and it is considered as a great driving force behind the current industrial revolution since it offers facile and feasible remedies for many problems. Because of the unique characteristic properties of nanomaterials, they can be employed in a wide variety of fields such as agriculture and food technology, catalysis, biomedical applications, tissue culture engineering, and fertilizers. In this regard, characterization of nanomaterials plays a significant role in determining their optical, thermal, and physicochemical properties. Many techniques have been used in nanomaterial characterization, and the most important techniques are discussed in detail in this book with their principles, basic operation procedures, and applications with suitable examples. In summary, this book offers broad content on the most important chemical and structural characterization techniques of nanomaterials.

The book offers comprehensive coverage of the most essential topics, including the following:

- Provides a comprehensive understanding of physical and chemical characterization techniques of nanomaterials
- Includes details about basic principles of each characterization technique with appropriate examples
- Covers most of the important characterization techniques that should be known to undergraduate/early career scientists/ beginners in materials chemistry

- Provides all the basic knowledge to understand and carry out the respective analysis of nanomaterials
- Fulfills the timely need of a book that covers the most important and useful characterization techniques in nanomaterial characterization

Up to date, there are no other books/book chapters which discuss most of these nanocharacterization techniques in one segment with all the basic instrumentation details and narrated examples of nanomaterial characterization. In a nutshell, this book will be a great asset to undergraduates/early career scientists/beginners of material science as it provides a comprehensive and complete understanding of most of the techniques used in nanocharacterization tools in a short time. Intended audience is based on science education while specifically focusing on undergraduates/graduate students/early scientists and beginners of chemistry, materials chemistry, and nanotechnology and nanoscience.

# Characterization Techniques for Nanomaterials

Dr. Imalka Munaweera
University of Sri Jayewardenepura, Sri Lanka
M.L. Chamalki Madhusha
University of Sri Jayewardenepura, Sri Lanka

**CRC Press**
Taylor & Francis Group
Boca Raton  London  New York

CRC Press is an imprint of the
Taylor & Francis Group, an **informa** business

First edition published 2023
by CRC Press
6000 Broken Sound Parkway NW, Suite 300, Boca Raton, FL 33487-2742

and by CRC Press
4 Park Square, Milton Park, Abingdon, Oxon, OX14 4RN

*CRC Press is an imprint of Taylor & Francis Group, LLC*
© 2023 Imalka Munaweera and M.L. Chamalki Madhusha

ISBN: 9781032406619 (hbk)
ISBN: 9781032406640 (pbk)
ISBN: 9781003354185 (ebk)

DOI: 10.1201/9781003354185

Typeset in Times New Roman
by codeMantra

# CONTENTS

# PREFACE

Tailoring nanostructures is central to materials development for any technological application. Nanomaterials include information on the atomic, mesoscopic, and microscopic length scales, and their tailoring is enabled by characterization, which relates synthesis and processing of nanomaterials to their structure, properties, and performance. A major misinterpretation of structure/composition/ functional groups in nanomaterials may result in incomplete characterization with inaccurate information about the material. During my career, I have seen most of the undergraduates and research students face this problem due to lack of knowledge. In this regard, my main aim in writing this book was to describe in outline, and to set into context, all the important topics within the nanomaterial characterization techniques. Since this covers a very wide area, we have pointed out the structural and chemical characterization techniques and described them with relevant examples. Up to date, there are no other books/book chapters which discuss all these nanocharacterization techniques in one segment with all the basic instrumentation details and narrated examples of nanomaterial characterization. I have gone into particular details about the topics that are not well covered elsewhere. I have emphasized concepts and theories, as we believe that a good understanding of these is of more use in the long term than specifics of current systems, services, and techniques. Furthermore, I have also emphasized the historical dimension, as we believe it is essential to understand where the discipline and its constituents came from, and why some things are as they are.

This approach is rooted in the literature, with copious references, presented at the end of each chapter for ease of access. My hope is that the content of this book will be sufficient to give a basic understanding of nanomaterial characterization and that readers will follow the references for details and examples of those aspects in which they are most interested. The abstracts and the key references, at the end of each chapter, are intended to convey the basic messages in a concise way.

Dr. Imalka Munaweera

# AUTHORS

**Dr. Imalka Munaweera** is a winner of the 2021 OWSD-Elsevier Foundation Award for Early-Career Women Scientists in the Developing World for her research in Chemistry, Mathematics, and Physics especially for her research contribution to the nanotechnology-related projects. Currently, she is a Senior Lecturer at the Department of Chemistry at the University of Sri Jayewardenepura in Sri Lanka. She has over 10+ years of teaching experience in Nanotechnology, Application of Nanotechnology, Inorganic Chemistry, Polymer Chemistry, and Instrumental Analysis. She has over 15+ years of research experience in the field of Nanotechnology, Inorganic Chemistry, and Materials Science, as well as in instrumentation (AFM, SEM-EDX, TEM-EDX, FTIR, Raman, DLS, SQUID, etc.). She obtained her PhD in Chemistry at The University of Texas at Dallas, USA in 2015. She has served as an Assistant Professor in Chemistry, A&M University, Prairie View, Texas, USA. She was a Postdoctoral Researcher at the University of Texas Southwestern Medical Center, Dallas, Texas, USA. She was also a Graduate Research/Teaching Assistant in the Department of Chemistry at the University of Texas at Dallas, Texas, USA, during her graduate studies. Further, she was a Scientist at the Sri Lanka Institute of Nanotechnology, Sri Lanka, in her early career. Her research interests are nanotechnology for drug delivery/pharmaceutical applications, agricultural applications, and water purification applications. Furthermore, she also pursues research toward development of nanomaterials from natural resources for various industrial applications. She has authored over 20 publications in indexed journals and is also an inventor of four

US-granted patents, two Sri Lankan-granted patents (licensed and commercialized), and an international patent related to nanoscience and nanotechnology-based research. Further, she is a recipient of many awards related to nanoscience and nanotechnology research. In addition, she is a principal investigator (PI) for a research grant which was awarded by World Academy of Sciences (TWAS) and a core PI for two more grants (NRC PPP and TWAS). Apart from being awarded many accolades in both the US and Sri Lanka, including scholarships, she has also taken part in paper review and many conference presentations and contributed to abstract awards as well. Some of her awards include the National Science & Technology Award in Sri Lanka in 2010 under the category of innovative advanced technologies with commercial potential that was awarded by the President of Sri Lanka and several graduate competition awards awarded by the American Chemical Society.

**Ms. Chamalki Madhusha** is a graduate research assistant, and she has obtained her BSc degree in Chemistry from the University of Sri Jayewardenepura, Sri Lanka. Currently, she is working toward a sustainable future through green chemistry and nanotechnology-based findings. Her research interests include materials chemistry, food chemistry, green chemistry, and nanotechnology with high-impact journal publications. She is engaged in generating new research ideas and devising feasible solutions to broadly relevant problems. She is an author of eight indexed publications, two local and international patents, and four international conference abstracts.

# 1

# INTRODUCTION TO NANOMATERIAL CHARACTERIZATION

Compared with their bulk equivalents, nanomaterials that have at least one dimension in the sub-nanometer to 10 nm range exhibit distinctive characteristics, such as an exponential rise in reactivity (Cao, 2004). The unique physicochemical characteristics of nanomaterials such as their size, shape, composition, and surface qualities have a significant impact on their activity and performance. Understanding the complicated links among a nanomaterial's structure, its characteristics, and respective applications can be improved with the help of physicochemical characterization of the nanomaterials (Li et al., 2015). A key use of nanotechnology is the accurate assessment of the properties of nanomaterials, which necessitates the use of cutting-edge methods that are sensitive to nanoscale dimensions (Lee et al., 2012). The advancement of nanoscience and the synthesis of nanomaterials have occurred simultaneously during the past few decades with the development of techniques for nanomaterial characterization (Rao, Mukherjee, & Reddy, 2017). The goal of this book is to introduce the key tools that can be used to characterize the chemical composition; particle size; shape; surface charge; and optical, magnetic, and mechanochemical properties of nanomaterials (Munaweera et al., 2018).

A major misinterpretation of structure/composition/functional groups in surface may result in incomplete characterization with inaccurate information about the material. A crucial first step toward a more precise and accurate understanding and/or prediction of the safety and therapeutic/diagnostic efficacies of nanoparticles (NPs) is the collection of reliable and valid data on their characterization using the appropriate techniques (Mahmoudi, 2021). The planned usage of any NPs/nanomaterial, which is of utmost significance, has

DOI: 10.1201/9781003354185-1

a considerable influence on the type and scope of the characterization procedures. For instance, according to the US Food and Drug Administration's (FDA) nanotechnology regulatory science research plan guidance documents, "We intend our regulatory approach to be adaptive and flexible and to take into consideration the specific characteristics and effects of nanomaterials in the particular biological context of each product and its intended use"(Mahmoudi, 2021). The results could be deceptive and lead to inaccurate conclusions regarding the safety, blood residency, and therapeutic efficacy of NPs without the use of the proper characterization procedures that are appropriate for the intended usage. Numerous new uses for NPs are being discovered as the area of nanotechnology developments (McNeil, 2005). This involves the possible application of NPs in many areas such as food technology, bio-medical applications, agriculture, energy storage devices, and catalysis applications etc. (Emerich & Thanos, 2003; Zang, 2011; Madhusha, Munaweera, Karunaratne, & Kottegoda, 2020; Madhusha et al., 2021; Amarasinghe, Madhusha, Munaweera, & Kottegoda, 2022). Owing to the extraordinary properties of nanomaterials such as high surface area to volume ratios, which cause an exponential rise in reactivity at the molecular level, nanoscale materials frequently exhibit features that are distinct from those of their bulk counterparts (Ealia & Saravanakumar, 2017). These features include electronic, optical, and chemical properties as well as vastly different mechanical properties for NPs (Khan, Saeed, & Khan, 2019). As a result of the academic interest and the potential technical applications of NPs in numerous domains, they can be the subject of considerable research. There are numerous techniques that can be used to create these nanostructures, including mechanical, chemical, and other approaches (Minnich, Dresselhaus, Ren, & Chen, 2009). More types and quantities of nanomaterials are being synthesized now than ten years ago, necessitating the creation of more accurate and reliable techniques for their characterization. However, these characterization techniques can be challenging due to the inherent challenges associated with the correct analysis of nanoscale materials, as opposed to the bulk materials (as an example very small sample size and insufficient quantity in some scenarios where laboratory-scale production is followed) (Kuhlbusch, Asbach, Fissan, Göhler, & Stintz, 2011). In addition, not every research team can easily access a wide variety of characterization facilities due to the multidisciplinary nature of nanoscience and nanotechnology. In reality, it is frequently required to characterize NPs more comprehensively, necessitating a thorough strategy that combines complementary methodologies (Jeevanandam, Barhoum, Chan, Dufresne, & Danquah, 2018). In this context, knowing the strengths and weaknesses of the various characterization techniques is desirable to determine whether, in some circumstances, employing just one or two of such techniques is sufficient to produce accurate details/information while investigating a particular parameter such as particle size (Roduner, 2006). The scientific community is aware that there might be some variations between the operation of analytical characterization methods for nanomaterials and their more "traditional" modes of application for more "conventional" (macroscopic) materials. The fields of nanoscience and nanotechnology are continually evolving (Mourdikoudis, Pallares, & Thanh, 2018).

In this book, we describe in detail about how various approaches can be used to characterize NPs. These methods can be combined, or they can be used exclusively for the research of a certain attribute (Keelan, 2002). We compare each of these methods, taking into account certain factors such as their ease of access, cost, precision, non-destructiveness, ease of use, and affinity for particular compositions or materials. Despite the large number of techniques offered here, each one is carefully examined. The particle size, shape, and crystal structure of the nanomaterials can be determined using microscopy-based techniques such as Transmission Electron Microscopy (TEM), High Resolution-Transmission Electron Microscopy (HR-TEM), and Atomic Force Microscopy (AFM). Other methods like magnetic methods are tailored for particular classes of materials. These methods include superconducting quantum interference device magnetometry (SQUID) as an example. Numerous other methods offer further details about the structure, elemental make-up, optical characteristics, and other general and more focused physical characteristics of the NPs. These methods include X-ray spectroscopy and scattering techniques. This book is divided into two main sections (chemical and physical characterization), each of which will present a variety of unique characterization methods for NPs with respect to the attributes under consideration. The sections are divided into various groups techniques as previously mentioned (Figure 1.1).

Size and shape are two of the primary factors investigated under the characterization of NPs. Additionally, the size distribution, level of aggregation, surface charge, and surface area can be quantified to a certain extent to assess the surface chemistry of NPs (Jiang,

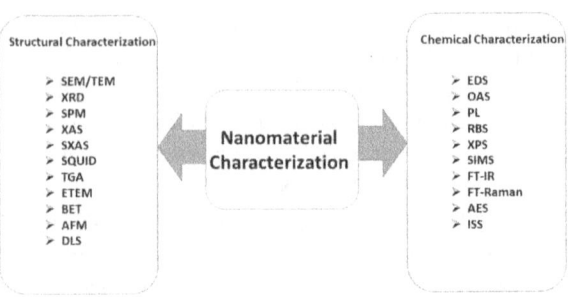

**Figure 1.1** An overview of characterization techniques of nanomaterials.

Oberdörster, & Biswas, 2009). Other characteristics and potential uses of the NPs may be influenced by size, size distribution, and organic ligands that are present on the surface of the particles. As a preliminary step after synthesis of NPs, the crystal structure and chemical makeup of the NPs are also carefully examined. For this objective, no established protocols have been reported. The use of these materials in commercial applications and the ability of the industry to adhere to regulations will both be significantly affected by the presence of reliable and trustworthy characterization tools for NPs. Significant challenges are associated with the analysis of nanomaterials due to the interdisciplinary nature of the field, the lack of appropriate reference materials for the calibration of analytical tools, the challenges associated with sample preparation for analysis, and the interpretation of the data. The measurement of in-situ NPs concentration, particularly in a scaling-up production system, and analysis in complicated matrices are additional unsolved problems in NPs characterization. Monitoring of waste and effluent from mass production of NPs is also necessary. More reliable measurement methods will be needed as the production of NPs is scaled up. Therefore, it is essential to fully characterize the nanomaterials created in various methods. Characterization is crucial to confirm whether the synthesized particles are at the nanoscale. Characterization in materials science refers to the general and comprehensive techniques used to investigate the properties and structure of the material (Chawla, 2012). For the subject to be understood scientifically, this fundamental procedure is necessary. Any process that deals with material analysis such as mechanical testing, thermal analysis, and density computation falls under the category of characterization because it involves tools necessary to examine material properties and microscopic structures (Vinson & Sierakowski, 2006). Characterization approaches that have been used for millennia are continuously being joined by newer and more sophisticated procedures (Lin, Lin, Wang, & Sridhar, 2014). Characterization enables us to evaluate the effectiveness of the procedure as well as the content and structure of the materials (Tantra, 2016). While some methods are quantitative, others are qualitative (Titus, Samuel, & Roopan, 2019).

## REFERENCES

Amarasinghe, T., Madhusha, C., Munaweera, I., & Kottegoda, N. (2022). Review on mechanisms of phosphate solubilization in rock phosphate fertilizer. *Communications in Soil Science and Plant Analysis, 53*(8), 944–960.

Cao, G. (2004). *Nanostructures & nanomaterials: synthesis, properties & applications*: Imperial college press.

Chawla, K. K. (2012). *Composite materials: science and engineering*: Springer Science & Business Media.

Ealia, S. A. M., & Saravanakumar, M. (2017). *A review on the classification, characterisation, synthesis of nanoparticles and their application.* Paper presented at the IOP conference series: materials science and engineering.

Emerich, D. F., & Thanos, C. G. (2003). Nanotechnology and medicine. *Expert Opinion on Biological Therapy, 3*(4), 655–663.

Jeevanandam, J., Barhoum, A., Chan, Y. S., Dufresne, A., & Danquah, M. K. (2018). Review on nanoparticles and nanostructured materials: history, sources, toxicity and regulations. *Beilstein Journal of Nanotechnology, 9*(1), 1050–1074.

Jiang, J., Oberdörster, G., & Biswas, P. (2009). Characterization of size, surface charge, and agglomeration state of nanoparticle dispersions for toxicological studies. *Journal of Nanoparticle Research, 11*(1), 77–89.

Keelan, B. (2002). *Handbook of image quality: characterization and prediction*: CRC Press.

Khan, I., Saeed, K., & Khan, I. (2019). Nanoparticles: properties, applications and toxicities. *Arabian Journal of Chemistry, 12*(7), 908–931.

Kuhlbusch, T. A., Asbach, C., Fissan, H., Göhler, D., & Stintz, M. (2011). Nanoparticle exposure at nanotechnology workplaces: a review. *Particle and Fibre Toxicology, 8*(1), 1–18.

Lee, D.-E., Koo, H., Sun, I.-C., Ryu, J. H., Kim, K., & Kwon, I. C. (2012). Multifunctional nanoparticles for multimodal imaging and theragnosis. *Chemical Society Reviews, 41*(7), 2656–2672.

Li, X., Liu, W., Sun, L., Aifantis, K. E., Yu, B., Fan, Y., Watari, F. (2015). Effects of physicochemical properties of nanomaterials on their toxicity. *Journal of Biomedical Materials Research Part A, 103*(7), 2499–2507.

Lin, P.-C., Lin, S., Wang, P. C., & Sridhar, R. (2014). Techniques for physicochemical characterization of nanomaterials. *Biotechnology Advances, 32*(4), 711–726.

Madhusha, C., Munaweera, I., Karunaratne, V., & Kottegoda, N. (2020). Facile mechanochemical approach to synthesizing edible food preservation coatings based on alginate/ascorbic acid-layered double hydroxide bio-nanohybrids. *Journal of Agricultural and Food Chemistry, 68*(33), 8962–8975.

Madhusha, C., Rajapaksha, K., Munaweera, I., de Silva, M., Perera, C., Wijesinghe, G., Kottegoda, N. (2021). A Novel green approach to synthesize curcuminoid-layered double hydroxide nanohybrids: adroit biomaterials for future antimicrobial applications. *ACS Omega, 6*(14), 9600–9608.

Mahmoudi, M. (2021). The need for robust characterization of nanomaterials for nanomedicine applications. *Nature Communications, 12*(1), 1–5.

McNeil, S. E. (2005). Nanotechnology for the biologist. *Journal of Leukocyte Biology, 78*(3), 585–594.

Minnich, A., Dresselhaus, M. S., Ren, Z., & Chen, G. (2009). Bulk nanostructured thermoelectric materials: current research and future prospects. *Energy & Environmental Science, 2*(5), 466–479.

Mourdikoudis, S., Pallares, R. M., & Thanh, N. T. (2018). Characterization techniques for nanoparticles: comparison and complementarity upon studying nanoparticle properties. *Nanoscale, 10*(27), 12871–12934.

Munaweera, I., Shaikh, S., Maples, D., Nigatu, A. S., Sethuraman, S. N., Ranjan, A., Chopra, R. (2018). Temperature-sensitive liposomal ciprofloxacin for the treatment of biofilm on infected metal implants using alternating magnetic fields. *International Journal of Hyperthermia, 34*(2), 189–200.

Rao, B. G., Mukherjee, D., & Reddy, B. M. (2017). Novel approaches for preparation of nanoparticles *Nanostructures for Novel Therapy* (pp. 1–36): Edited by Denisa Ficai and Alexandru Mihai Grumezescu: Elsevier.

Roduner, E. (2006). Size matters: why nanomaterials are different. *Chemical Society Reviews, 35*(7), 583–592.

Tantra, R. (2016). *Nanomaterial characterization: An introduction*: John Wiley & Sons.

Titus, D., Samuel, E. J. J., & Roopan, S. M. (2019). Nanoparticle characterization techniques *Green synthesis, characterization and applications of nanoparticles* (pp. 303–319): Edited by Ashutosh Kumar Shukla and Siavash Iravani: Elsevier.

Vinson, J. R., & Sierakowski, R. L. (2006). *The behavior of structures composed of composite materials* (Vol. 105): Springer.

Zang, L. (2011). *Energy efficiency and renewable energy through nanotechnology*: Springer.

# 2

# CHEMICAL AND STRUCTURAL CHARACTERIZATION OF NANOMATERIALS

Elucidation of molecular structure of a chemical substance is necessary to identify or confirm the structural identity of a chemical compound during chemical research or product developments. Therefore, one must be aware that several of the parameters that need to be established when using NPs in test systems depend significantly on the environment and the temporal evolution of the nanomaterials. Therefore, if free NPs are present in the formulation supplied to the end-user or the environment, it is important to evaluate the nanomaterials in that formulation exactly as they were manufactured. Nanomaterials can take the form of nanopowders, ultrafine particles, NPs, aerosols, colloids suspended in liquids and solids with embedded nanomaterials, etc. Manufactured nanomaterials need to be disseminated in the proper media in order to assess their biological safety. The behavior of the suspension may be significantly affected by the interaction between these media and the nanomaterials (Bhagyaraj & Oluwafemi, 2018).

The significance of the probable dissolution kinetics needs to be emphasized with the growing number of recently developed manmade nanomaterials. Nanomaterials are likely to dissolve significantly faster than their respective bulk materials since the kinetics of dissolution are typically inversely proportional to the surface area. This is true, for instance, with silver NPs, which are increasingly used as anti-bacterial agents, due to their release of silver ions. However, the kinetics of the aforementioned dissolution has not yet been thoroughly investigated. The complexity of risk assessments for nanomaterials is highlighted by the example of silver NPs since it is necessary to distinguish between the negative interactions of silver NPs with biological systems and those interactions with ionic

DOI: 10.1201/9781003354185-2

silver (Reidy, Haase, Luch, Dawson, & Lynch, 2013). It is significant to note that not every property can be identified under every circumstance nor is it required to do so.

The number, form, and topological organization of phases and defects such as point defects, dislocations, stacking faults, or grain boundaries in a crystalline material characterize the microstructure. Polycrystalline solids having grain sizes less than 100 nm are known as nanocrystalline materials. Chemical and physical scale effects result from this nano-specific microstructure (nanostructure). Understanding the properties of nanomaterials necessitates a comprehensive understanding of structure from the atomic/molecular (local) to the crystal structure (long-range order) and to the microstructure (mesoscopic scale and defect structure). To characterize the nanomaterials at all length scales, a variety of analytical approaches are needed (Meyers, Mishra, & Benson, 2006).

## 2.1 PARTICLE MORPHOLOGY AND SIZE

The average particle and grain size, their distribution, and the morphology of the nanoparticles are all determined using scattering and imaging techniques such as transmission electron microscopy, small angle X-ray and neutron scattering, and line broadening in X-ray diffraction. Additional information on these size-related characteristics is provided via nitrogen adsorption-desorption, light scattering techniques, atomic probing methods, and mass spectroscopic methods. The method chosen will vary depending on the nature of the material such as powder, solid, and liquid dispersion, and consolidated and sintered ceramic. Due to their large surface area and nanoscale size, NPs have distinct physical and chemical characteristics. According to the reports, the size of NPs influences their optical characteristics and imparts various colors through visible-range absorption. Their distinctive size, shape, and structure also affect their reactivity, toughness, and other qualities. In this regard, information on nanoparticle morphology and size plays a vital role in nanomaterial characterization.

## 2.2 LOCAL STRUCTURE

Diffuse scattering also contains information about the local structure as do spectroscopic techniques like nuclear magnetic resonance

(NMR) or extended X-ray absorption fine structure (EXAFS) spectroscopy. A significant portion of the atoms in heterogeneously disordered nanocrystalline materials is found at surfaces and interfaces. It has been demonstrated that atoms near particle surfaces, which have greater degrees of freedom than the atoms in the core of the particles, exhibit higher levels of disorder in pure tetragonal nanocrystalline zirconia with a particle size of 5 nm (Tsunekawa, Ito, Kawazoe, & Wang, 2003). Local molecular processes play a major role in the production of particles by chemical vapor synthesis (CVS) and the evolution of the microstructure during sintering. Reverse Monte Carlo simulation of EXAFS spectra revealed that segregation of aluminum atoms is the mechanism for grain growth suppression during sintering of zirconia doped with alumina. Similarly, in zirconia, an inhomogeneous distribution of yttrium was discovered and linked to reduced sinterability (Horvath, Birringer, & Gleiter, 1987).

## 2.3 CRYSTAL STRUCTURE

X-ray diffraction, electron diffraction/scattering, and neutron diffraction techniques are often utilized to characterize the crystal structure, which have been discovered to be frequently size-dependent. The precision of crystal structure identification is limited by the line broadening at very small crystallite sizes which makes it challenging to differentiate between tetragonal and cubic zirconia. In general, it is seen that all diffraction reflexes are present and that the background is extremely low for CVS nanopowders. This suggests that there are many crystals and few defects inside the individual particles. One exception is nanocrystalline silicon carbide, which, when particularly manufactured at low temperature, exhibits stacking defects and twinning. Rietveld refinement not only provides the phase composition, lattice parameters, and positions of atoms in the unit cell but also the crystallite size and the microstrain can be extracted (Tsunekawa et al., 2003).

## 2.4 MICROSTRUCTURE

The electron microscopy (both scanning and transmission) is heavily utilized to characterize flaws like dislocations, grain boundaries, agglomerate size and structure, distribution of phases, grain and phase boundaries, pores, etc., that are crucial for the properties of NPs. The main NPs in CVS nanopowders are single crystal and have lattice fringes that reach the surface. Additionally, crystallographic habitus planes are typically observed in the NPs. Nitrogen adsorption/desorption and small angle scattering are employed to quantify specific surface area, total pore volume, pore diameter, and pore size distributions for the purpose of characterizing porosity in NPs. Powders display the fractal patterns commonly seen in aerogels (Steacy & Sammis, 1991).

## REFERENCES

Bhagyaraj, S. M., & Oluwafemi, O. S. (2018). Nanotechnology: the science of the invisible *Synthesis of inorganic nanomaterials* (pp. 1–18): Elsevier.

Horvath, J., Birringer, R., & Gleiter, H. (1987). Diffusion in nanocrystalline material. *Solid State Communications, 62*(5), 319–322.

Meyers, M. A., Mishra, A., & Benson, D. J. (2006). Mechanical properties of nanocrystalline materials. *Progress in Materials Science, 51*(4), 427–556.

Reidy, B., Haase, A., Luch, A., Dawson, K. A., & Lynch, I. (2013). Mechanisms of silver nanoparticle release, transformation and toxicity: a critical review of current knowledge and recommendations for future studies and applications. *Materials, 6*(6), 2295–2350.

Steacy, S. J., & Sammis, C. G. (1991). An automaton for fractal patterns of fragmentation. *Nature, 353*(6341), 250–252.

Tsunekawa, S., Ito, S., Kawazoe, Y., & Wang, J.-T. (2003). Critical size of the phase transition from cubic to tetragonal in pure zirconia nanoparticles. *Nano Letters, 3*(7), 871–875.

# 3

# PARAMETERS OF NANOMATERIALS WHICH SHOULD BE CHARACTERIZED

## 3.1 MORPHOLOGICAL FEATURES

The morphological characterization of NPs has a paramount signifi-cance because the morphology of nanomaterials profoundly affects majority of the nanomaterials' properties (Bhagyaraj & Oluwafemi, 2018). Morphological features of different nanomaterials vary sig-nificantly depending on lattice and crystal structure, matrix composi-tion, synthesis methodology, and thermal stability or photo stability. Furthermore, morphological variation is a facile and effective way of introducing specific functionality to nanomaterials by affecting their biocompatibility as well (Kumar, Depan, Tomer, & Singh, 2009). As an example, a self-assembling duplex DNA has been used as build-ing blocks to synthesize 3- dimensional DNA structures between 10 nm and 100 nm (Xu et al., 2019). A specific technique was used to create a different nanoscale "DNA origami" which possesses a dif-ferent morphology than the used 3-dimensional DNA structure (Shin, Yuk, Chun, Lim, & Um, 2020). Therefore, morphological diversity is very significant in determining the functionalities of nanoma-terials. Nanomaterials are available in a wide variety of forms and each one of them is distinctive. As an illustration, consider spherical nanospheres, nanoreefs, nanocorals, nanoboxes, nanotubes, nanoclus-ters, nanocoils, nanoballs, and nanocones, among other structures (Burda, Chen, Narayanan, & El-Sayed, 2005) (Figure 3.1). These shapes appear as a result of two processes: the spontaneous evolution of shapes during the synthesis process, and the effects of a templat-ing or guiding agent. As an example, formation of micellar emulsions can be mentioned (Glatter & Salentinig, 2020). The other factor that determines the aforementioned morphologies is the innate crystallo-graphic growth patterns of the nanomaterials themselves. Amorphous

    DOI: 10.1201/9781003354185-3

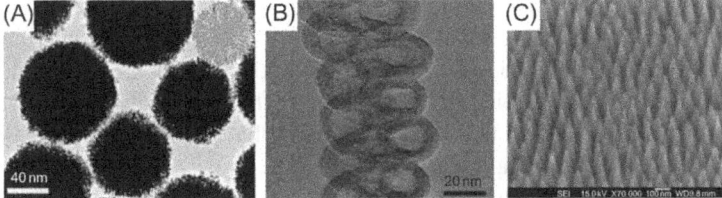

**Figure 3.1** TEM images of different kinds of 3D nanomaterials (A) nanoballs (dendritic structures), (B) nanocoils, and (C) nanocones (Mayeen, Shaji, Nair, & Kalarikkal, 2018).

particles usually adopt spherical shapes or nanospheres and anisotropic microcrystalline whiskers which correspond to their particular crystal shape. Small NPs usually form clusters, and these may be of a variety of shapes like corals, triangles, rods, fibers, and plates. The study of fine particles is called micromeritics (Sayes & Warheit, 2009).

The control over the morphology of nanomaterials is of utmost importance to effectively explore the features of nanomaterials for usage in a variety of future technologies and applications. Examples of applications based on the optical properties of gold NPs include optical filters and bio-sensors (Sheikhzadeh, Beni, & Zourob, 2021). These applications necessitate anisotropy of the particle shape because bigger shapes result in higher plasmon losses (Li, Zhao, & Astruc, 2014). Since well-defined magnetization axis and switching fields are necessary to process or store information, the shape of magnetic NPs cannot be regulated (Gubin, 2009). Furthermore, to learn more about the morphological characteristics of NPs, flatness, sphericity, and aspect ratio will be considered in nanomaterials (Sanjay & Pandey, 2017).

Nanowires and nanotubes with various shapes such as belts, helices, zigzags, or nanowires with diameter that varies with length are examples of NPs with high aspect ratios. However, spherical, oval, cubic, prismatic, helical, or pillar-shaped NPs have smaller aspect ratios (Buzea, Pacheco, & Robbie, 2007). These NPs can be found as colloids, suspensions, or solid powders. Additionally, the concept of aspect ratio is the foundation in categorizing nanomaterials based on their dimensionality. In the nanoscale, 1D nanomaterials are one-dimensional. These are made up of surface coatings or thin films. Two dimensions are present in 2D nanomaterials at the nanoscale

scale. These include nanopore filters or 2D nanostructured films with nanostructures securely affixed to a substrate. Materials that are of three-dimensional nanoscale are known as 3D nanomaterials. These include colloids, free NPs with different morphologies, and thin films produced under circumstances that result in atomic-scale porosity(Paramasivam et al., 2021).

There are many methods for characterizing nanomaterials to examine their morphology, but the most important ones include the use of microscopy (Pal, Jana, Manna, Mohanta, & Manavalan, 2011). These include scanning electron microscopy (SEM), transmission electron microscopy (TEM), and polarized optical microscopy (POM) (Yang, Coombs, & Ozin, 1997). The SEM is based on the electron examining rule and provides all available information at the nanoscale level about the nanomaterials including the identification of their morphological properties as well as how NPs have scattered within the lattice structure. The Single Walled Nanotubes (SWNTs) that are dispersed in the nylon-6 and poly(butylene) terephthalate (PBT) polymer matrix were examined using the SEM technique (Coleman, Khan, & Gun'ko, 2006). Additionally, the POM approach has shown that nanomaterials feature spherulites that resemble stars (Jing, Shi, Zhang, & Qin, 2015). The larger loading of SWNTs in the lattice matrix resulted in smaller sizes for the nanomaterials. Additionally, the SEM technique was used to examine the morphological characteristics of ZnO-doped molecular orbital frameworks (Huang et al., 2019). This approach exposes the morphologies of molecular orbital frameworks and has proven the scattering of ZnO NPs in the lattice. Since TEM is reliant on the electron transmission standard, it can reveal several distinctive information about the bulk content when exposed to currents of low to high amplification. The morphologies of gold NPs, for instance, have been studied using TEM (Ong, Luo, & Stellacci, 2017). TEM images of gold NPs have shown certain NP properties (Dykman & Khlebtsov, 2011). TEM also provides fundamental information about nanomaterials with layered structures, such as in the $Co_3O_4$ NPs. In these NPs, the empty quadrupolar shell structure is clearly made visible by TEM (Afolalu, Soetan, Ongbali, Abioye, & Oni, 2019).

## 3.2 STRUCTURAL DEFECTS

For the characterization of structural deficiencies in nanomaterials, numerous characterization approaches have been developed. Direct and indirect methods can be used to characterize these procedures.

The microscopic methods, such as SEM or TEM, are included in the category of direct procedures. The investigation of grain boundaries using electron backscatter diffraction (EBSD) in the context of the SEM is precise and successful. When it comes to indirect methods, the main techniques that are frequently utilized include electrical resistometry (ER), X-ray line profile analysis (XLPA), and positron annihilation spectroscopy (PAS) (Gubicza, 2017). Characterization is straightforward and simple with techniques like PAS, ER, and XLPA since they are nondestructive techniques. Thin surface layers can be analyzed by utilizing the PAS and XLPA. In this case, the surface of the relevant specimen must be chopped or machined and removed using electropolishing or etching. In addition, mechanical polishing of nanomaterials is not advised as a suitable surface preparation for PAS or XLPA since it may seriously alter the nanomaterial's surface layer. As a result, the results are connected to the surface fault structure that has been mechanically polished and do not characterize the bulk material. Therefore, this outermost deformed layer must be removed prior to PAS or XLPA analyses if earlier investigations such as hardness assessment required a mechanically polished surface. EBSD stands out when looking at non-destructive procedures. The most accurate and sensitive method for determining the surface quality of nanomaterials is EBSD. This is brought about by the extremely thin surface layer analyzed in this technique and the method's susceptibility to contamination and surface roughness (Gubicza, 2017). In general, EBSD technology requires a pure, smooth, and distortion-free surface. As a result, there are a number of procedures frequently involved in surface preparation of the particular nanomaterial which are briefly covered here. Mechanical polishing is the first step toward obtaining a smooth surface. Then, electropolishing or ion polishing is required to remove the mechanical polishing-induced distortion in the topmost layer. The most challenging sample preparation is required for TEM, which includes mechanical thinning of the sample to a thickness of 80 mm and then jet polishing or ion milling the sample until it is perforated (Mukhopadhyay, 2003). The sample is frequently chilled to liquid nitrogen temperature (77 K) during the thinning by an ion beam in order to prevent the recovery and recrystallization of nanostructures due to the temperature rise produced by the ion beam. Compared to TEM, ER method examines a bigger volume. Additionally, the volume probed by XLPA or PAS is considerably bigger than that by TEM (Čížek et al., 2019). When comparing and discussing the results produced by various methodologies, these variations should be taken into account. The larger the analyzed volume of nanomaterial, the better the statistics of the parameters acquired for the defect structures.

## 3.3 OPTICAL STUDIES

One of the major characteristic properties that govern the photocatalytic application of nanomaterials is their optical property. When considering the optical properties of nanomaterials, Beer Lambert law plays a significant role and it states the quantity of light absorbed by a substance dissolved in a fully transmitting solvent is directly proportional to the concentration of the substance and the path length of the light through the solution. The techniques that are used in optical characterization are important to reveal the details of light absorption, reflectivity of nanomaterials, luminescence, and phosphorescence properties. Semiconductor nanomaterials and metallic NPs, for example, are extensively used for photographic applications (Khan, Saeed, & Khan, 2019).

The instruments that are being used in identification of optical properties are UV-Vis spectrophotometer, photoluminescence, and spectroscopic ellipsometry. The Diffuse Reflectance Spectrometer (DRS) is mainly used for the evaluation of band gap values and solid UV spectra of nanomaterials (Sangiorgi, Aversa, Tatti, Verucchi, & Sanson, 2017). DRS technique is specific since it is used for only solid samples to measure the band gaps.

Photo activity and conductivity of nanomaterials are crucial for bandgap evaluation. Graphite-carbon nitride ($C_3N_4$) is considered as the photocatalyst for the metal-free water part. The photographic limits of this nanomaterial are explicitly linked to the UV-Vis spectroscopy (2.74–2.77 eV) bandgap estimate. In addition, this framework also provides absorption movement in case of doping event, compound action, or hetero-structure of nanomaterials. To examine the range of optical properties of $LaFeO_3$, montmorillonite, and $LaFeO_3$/montmorillonite nanocomposites, their UV electromagnet absorption range was investigated (Afolalu et al., 2019). The existence of the nanocomposite when the montmorillonite and $LaFeO_3$ NPs were in their purest form was confirmed by a dramatic red shift. $LaFeO_3$/montmorillonite and $LaFeO_3$ demonstrated a comparatively wide range of osmosis between 400 nm and 620 nm. Photoluminescence (PL) is also considered a fundamental technology for investigating the optical properties of photoactive NPs despite their UV effects (Li & Zhang, 2009). The retention or breakdown of the material's radiation point and its impact on the overall photo excitons of the excitation time are further explained by this method. As a result, it offers a wealth of knowledge on load reattachment and the partial presence of reinforced materials in their front band which is helpful for all imaging and photography applications. The PL range may be registered as a release depending on the study. A typical PL range is developed with defect-free and balanced ZnO NPs, and from this range it is clear that the defect-free ZnO NPs have higher PL control when the ZnO NPs changed to CdS (Eixenberger et al., 2019). The CdS/Au/ZnO composite shows the lowest performance. In the recent past, this damping drop from CdS/Au/ZnO to unaltered ZnO can be attributed to a decrease in the charge recombination rate and a longer photo exciton lifetime (Gurugubelli, Ravikumar, & Koutavarapu, 2022). Additionally, the doping rate of the material in the NPs, layer thickness, and oxygen opening are all determined using this framework. Wan et al. also used spectroscopic ellipsometry methods to select estimates of refractor reduction and tail coefficients for bare gold NPs (Beaudette, 2021). To identify optical constants with various morphologies and plasmonic properties, they organized the motion of gold NPs. The properties from the optical continuum estimates of solid gold NPs exhibit outstanding indications of utilizing these materials in compound identification applications given their sensitivity as indicated by ellipsometric values (Beaudette, 2021).

## 3.4 STRUCTURAL ANALYSIS

Structural analysis of nanomaterials is significant since it is a prerequisite for better understanding of the characteristics of nanomaterials to have a detailed knowledge of the structure from the atomic/molecular (local) level to the crystal lattice structure and to the microstructure. The main objectives of structural analysis of nanomaterials are to investigate the relationship between structure and property as well as to reveal novel properties.

A key method for determining the size and crystallite structure of nanomaterials is powder X-ray diffraction (PXRD). The morphological and structural details of the researched nanomaterials can be studied using X-ray scattering and Bragg diffraction (Giannini et al., 2016). From the PXRD analysis, we can deduce a variety of structural details including the atomic structure of the crystal, the positions and symmetry of the atoms in the unit cells, the size and shape of the nanocrystalline domain, the identification of the crystalline phases and a quantitative estimation of their weight fractions and the positions and symmetry of the nanoscale assembly's NPs and nanocrystals as well as the assembly's length (Hens & De Roo, 2020). Additionally, TEM provides a clear, consistent, and straightforward technique to visualize the atomic lattice in a crystalline nanomaterial (Chen et al., 2013). The method's strength and significance are further increased by the ability to reveal minute crystal structure features. However, a major drawback of the traditional TEM is that the imaging resolution that is theoretically predicted cannot be achieved (Wall & Hainfeld, 1986). The reasons behind this drawback are the inability to control the image forming electron beam in the instrument and stability of the high voltage source for the filament or level of vacuum. As a result of the aforementioned reasons, true atomic resolution images have been incomprehensible. With the development of nanoscience, new advances in TEM technology have increased the resolution limit to the sub-angstrom scale (Ercius, Alaidi, Rames, & Ren, 2015). By using aberration-corrected scanning transmission electron microscopy (STEM) which relies on rastering a focused electron beam over the sample and a high angle annular dark field (HAADF) detector which provides enhanced Z contrast, electron microscopy images with an atomic resolution (nominally 0.8 Å) have been achieved (Sang et al., 2017). Such capability opens the door to the investigation of the atomic level structure of nanomaterials which may have a significant impact on their properties. These analysis methods enabled elemental mapping, Z contrast analysis, and atomic scale imaging (with an EDS instrument attachment) (Wirth, 2009). The findings have given us a unique perspective on the intricate structure of NPs and distinctive proof that sophisticated nanomaterials exist.

## 3.5 ELEMENTAL STUDIES

There are various kinds of medicinal, commercial, and bio-based applications of nanomaterials in the recent era. Regardless of the infinitesimally small size of nanomaterials, they can pose significant risks to both humans and their health. In this regard, a detailed and complete analysis of the elemental composition of respective nanomaterials is vital to ensure the safe and sustainable establishment of nanomaterials in human applications. The most trusted and well-known method to analyze the elemental composition of nanomaterials is Mass Spectrometry (MS) (Chait, 2011). MS determines the mass-to-charge ratio of ions in the interfacial layer of respective nanomaterial and is used to determine atoms or molecules in the nanomaterial (Domon & Aebersold, 2006).

When performing MS measurements, various techniques have been employed to ionize the nanomaterial sample which results in the molecules gaining a charge and finally resulting in fragmentation. The resulted fragments are then subjected to mass selection, which separates those nanomaterials according to their respective mass-to-charge ratio. Then, separate groups of molecules can be analyzed to provide more details about the elemental composition of the nanomaterial. With the development of nanotechnology, researchers have investigated two methods of single-particle inductively coupled plasma mass spectrometry (spICP-MS) (Mitrano et al., 2014). ICP technique creates ions to interact with the nanomaterial and involves the generation of electrically neutral plasma from argon gas (Banoub et al., 2015). This method has been tested by using the time-of-flight (TOF) and quadrupole (Q) analyzers (Kristensen, Imamura, Miyamoto, & Yoshizato, 2000).

The time taken by NPs (after acceleration by an electric field) to reach the detector will be measured by TOF. In Q analyzers, four parallel rods form a radio frequency quadrupole field (one sequence of electrical current), which effectively filters various elements via the instrument. It has been shown that both mass spectrometry methods are useful for determining the elemental makeup of individual NPs. However, it has been found that the TOF method outperformed the Q method at detecting smaller NPs. The optimal method for multi-element detection of nanomaterials is hence spICP-TOFMS (Praetorius et al., 2017).

Elemental mapping utilizes the compositional accuracy built into methods like EDS microanalysis and combines it with high-resolution imaging to present complex data in an approachable, aesthetically appealing format. The foundation of elemental mapping is the collection of incredibly precise elemental composition data over a sample's surface. Usually, a SEM or TEM is used for this, along with EDS analysis. Along with the EDS data, a high-resolution image of the area of interest is gathered, and the two are then correlated. A full elemental spectrum is also gathered for every pixel in the digital image.

## 3.6 SIZE ESTIMATION

The size estimation of NPs can be characterized by different techniques which include TEM, SEM, Atomic Force Microscopy (AFM), X-ray Diffraction (XRD), and Dynamic Light Dispersing (DLS). Usually, TEM, AFM, XRD, and SEM techniques give better estimation of the size of nanomaterials than DLS measurements because the latter technique estimates the size of nanomaterials at incredibly low dimensions. The DLS technique has been used to study the size variation of silica NPs with the absorption rate of serum protein. The size of the silica NPs has expanded with the protein layer. In this regard, the hydrophilicity of respective NPs and agglomeration of NPs may affect the DLS measurements (Mourdikoudis, Pallares, & Thanh, 2018). Therefore, the Differential Centrifugal Sedimentation (DCS) technique is considered in such situations (Langevin et al., 2018). For the investigation of DNA, proteins, and other organic compounds, a rather strong and distinctive methodology is available and it helps to track NPs in addition to the DCS method (Yang, Lu, Fan, & Murphy, 2014). Furthermore, the size distribution profile of the NP in a fluid medium with diameters ranging from 10 to 1,000 nm can be located by correlating the rate of Brownian motion to the particle size. This is done by using the NP tracking analysis (NTA) strategy to analyze and visualize NPs in the liquid medium (Afolalu et al., 2019). Compared to DLS measurements, the NTA technique interpret more accurate and precise results with far better peak resolution when used in sizing monodispersed and poly-dispersed samples (OKTAVIANI, 2021).

The large surface area of nanomaterials provides exceptional and extraordinary performance for wide variety of applications while Brunauer-Emmett-Teller (BET) method is the most widely used technique to measure the surface area. This technique is based upon the Brunauer-Emmett-Teller (BET) hypothesis and the adsorption and desorption rules (Tian & Wu, 2018).

## 3.7 PHYSICOCHEMICAL CHARACTERISTICS

Physicochemical properties of nanomaterials include size, shape of the nanomaterials, solubility in different solvent systems, surface area measurements, chemical composition, shape, agglomeration state, crystallinity, surface energy, surface charge, surface morphology, and surface coating (Gatoo et al., 2014). Moreover, the electronic and optical characteristics of nanomaterials play a major role in most of the applications hence the characterization of the physicochemical properties of nanomaterials is essential. As an example, metallic nanomaterials exhibit collective oscillation bands of electrons excited by the incident photons at the resonant frequency and we can clearly see these bands by a UV spectrum. However, these oscillation bands are apparently absent in mass metal scope and such observations can be exploited by localized surface plasma resonance (LSPR) (Singh & Strouse, 2010). The excitation of LSPR bands enables the tunable and enhanced electromagnetic fields, light absorption and scattering based on the physical and elemental parameters of nanomaterials which in turn upgrades the performances. The LSPR spectrum interprets the shape, dimensions and interparticle which differentiates the nanomaterials from dielectric properties including solvents, adsorbents, and substrates (Gatoo et al., 2014). As an example, the mean free path for silver metal is roughly 50 nm. The LSPR spectrum of silver shows three distinct bands corresponding to the in-plane dipole, quadrupole, and out-plane quadrupole plasmon resonance (Wu, Zhou, & Wei, 2015). The SPR shifted to shorter wavelengths and as a result of that, the edge length of the silver NPs has decreased.

## 3.8 MAGNETIC PROPERTIES

Magnetic properties of nanomaterials have paramount importance in most of the applications in modern nanotechnology. Therefore, the estimation of magnetic properties of nanomaterials is vital (Kolhatkar, Jamison, Litvinov, Willson, & Lee, 2013). The magnetic characteristics of NPs generally rely on size, and nanomaterials can be used for the synthesis and design of materials with desired magnetic properties based on the size factor. Nanomaterials are made up of large magnetic particles that are typically divided into a number of magnetic domains with distinct magnetization directions. Small magnetic particles in certain nanomaterials with dimensions less than a threshold size, however, are made up of a single domain. Magnetization in single-domain particles may be stable, however magnetization in very small particles is unstable above the superparamagnetic blocking temperature. (Dunlop, 1973) This is due to the fact that the energy required to reverse the magnetization is proportional to the volume, and the thermal energy may then be sufficient to result in superparamagnetic relaxation, such as changes in the magnetization direction of extremely small particles. Superparamagnetic relaxation renders the particles unsuitable for magnetic data storage applications, but it may be required at maximum performance in many other applications, such as magnetic beads used in biotechnology. (Houshiar, Zebhi, Razi, Alidoust, & Askari, 2014) Superparamagnetic relaxation can be explored using a variety of timescale-dependent techniques, including dc and ac magnetization measurements, Mossbauer spectroscopy, and neutron scattering. (Mørup, Madsen, Frandsen, Bahl, & Hansen, 2007) As a result, these techniques can be utilized to investigate superparamagnetic relaxation over a wide variety of relaxation times. Magnetic fields can be used to suppress superparamagnetic relaxation. Strong interparticle interactions can potentially cause superparamagnetic relaxation to be suppressed. When considering magnetic dynamics, it is dominated by excitation of the uniform spin-wave mode, which exhibits a linear relationship between temperature and magnetization. (Fazlali et al., 2016) More importantly, the surface effects also dominate with the variation of magnetic properties of nanomaterials. With the increment of temperature, the surface magnetization decreases rapidly. As a result, magnetization in the interior of a particle will be lower than its surface magnetization. The magnetic anisotropy of nanomaterials will be prominently governed by the low symmetry around surface atoms. Moreover, the magnetic properties in the nanomaterial surface and the defects in the interior can be profoundly influenced by a reduced number of magnetic neighbor atoms and ultimately this can lead to non-collinear spin structures in ferrimagnetic particles. (Issa, Obaidat, Albiss, & Haik, 2013)

## REFERENCES

Afolalu, S., Soetan, S., Ongbali, S., Abioye, A., & Oni, A. (2019). *Morphological characterization and physio-chemical properties of nanoparticle-review.* Paper presented at the IOP Conference Series: Materials Science and Engineering.

Banoub, J., Delmas Jr, G. H., Joly, N., Mackenzie, G., Cachet, N., Benjelloun-Mlayah, B., & Delmas, M. (2015). A critique on the structural analysis of lignins and application of novel tandem mass spectrometric strategies to determine lignin sequencing. *Journal of Mass Spectrometry, 50*(1), 5–48.

Beaudette, C. A. (2021). *Nanostructures, nanoparticles, and 2D materials from nonthermal plasmas:* University of Minnesota.

Bhagyaraj, S. M., & Oluwafemi, O. S. (2018). Nanotechnology: the science of the invisible *Synthesis of Inorganic Nanomaterials* (pp. 1–18): Elsevier.

Burda, C., Chen, X., Narayanan, R., & El-Sayed, M. A. (2005). Chemistry and properties of nanocrystals of different shapes. *Chemical Reviews, 105*(4), 1025–1102.

Buzea, C., Pacheco, I. I., & Robbie, K. (2007). Nanomaterials and nanoparticles: sources and toxicity. *Biointerphases, 2*(4), MR17–MR71.

Chait, B. T. (2011). Mass spectrometry in the postgenomic era. *Annual Review of Biochemistry, 80*(1), 239–246.

Chen, C.-C., Zhu, C., White, E. R., Chiu, C.-Y., Scott, M., Regan, B., . . . Miao, J. (2013). Three-dimensional imaging of dislocations in a nanoparticle at atomic resolution. *Nature, 496*(7443), 74–77.

Čížek, J., Janeček, M., Vlasák, T., Smola, B., Melikhova, O., RK, I., & SV, D. (2019). The development of vacancies during severe plastic deformation. *Materials Transactions, 60*(8), 1533–1542.

Coleman, J. N., Khan, U., & Gun'ko, Y. K. (2006). Mechanical reinforcement of polymers using carbon nanotubes. *Advanced Materials, 18*(6), 689–706.

Domon, B., & Aebersold, R. (2006). Mass spectrometry and protein analysis. *Science, 312*(5771), 212–217.

Dunlop, D. (1973). Superparamagnetic and single-domain threshold sizes in magnetite. *Journal of Geophysical Research, 78*(11), 1780–1793.

Dykman, L., & Khlebtsov, N. (2011). Gold nanoparticles in biology and medicine: recent advances and prospects. *Acta Naturae (англоязычная версия), 3*(2 (9)), 36–38.

Eixenberger, J. E., Anders, C. B., Wada, K., Reddy, K. M., Brown, R. J., Moreno-Ramirez, J., Fologea, D. (2019). Defect engineering of ZnO nanoparticles for bioimaging applications. *ACS Applied Materials & Interfaces, 11*(28), 24933–24944.

Ercius, P., Alaidi, O., Rames, M. J., & Ren, G. (2015). Electron tomography: a three-dimensional analytic tool for hard and soft materials research. *Advanced Materials, 27*(38), 5638–5663.

Fazlali, M., Dvornik, M., Iacocca, E., Dürrenfeld, P., Haidar, M., Åkerman, J., & Dumas, R. K. (2016). Homodyne-detected ferromagnetic resonance of in-plane magnetized nanocontacts: Composite spin-wave resonances and their excitation mechanism. *Physical Review B, 93*(13), 134427.

Gatoo, M. A., Naseem, S., Arfat, M. Y., Mahmood Dar, A., Qasim, K., & Zubair, S. (2014). Physicochemical properties of nanomaterials: implication in associated toxic manifestations. *BioMed Research International, 2014*, Article ID 498420.

Giannini, C., Ladisa, M., Altamura, D., Siliqi, D., Sibillano, T., & De Caro, L. (2016). X-ray diffraction: a powerful technique for the multiple-length-scale structural analysis of nanomaterials. *Crystals, 6*(8), 87.

Glatter, O., & Salentinig, S. (2020). Inverting structures: from micelles via emulsions to internally self-assembled water and oil continuous nanocarriers. *Current Opinion in Colloid & Interface Science, 49*, 82–93.

Gubicza, J. (2017). *Defect structure and properties of nanomaterials*: Woodhead Publishing.

Gubin, S. P. (2009). *Magnetic nanoparticles*: John Wiley & Sons.

Gurugubelli, T. R., Ravikumar, R., & Koutavarapu, R. (2022). Enhanced photocatalytic activity of ZnO–CdS composite nanostructures towards the degradation of rhodamine B under solar light. *Catalysts, 12*(1), 84.

Hens, Z., & De Roo, J. (2020). Atomically precise nanocrystals. *Journal of the American Chemical Society, 142*(37), 15627–15637.

Houshiar, M., Zebhi, F., Razi, Z. J., Alidoust, A., & Askari, Z. (2014). Synthesis of cobalt ferrite (CoFe2O4) nanoparticles using combustion, coprecipitation, and precipitation methods: a comparison study of size, structural, and magnetic properties. *Journal of Magnetism and Magnetic Materials, 371*, 43–48.

Huang, L., Yang, Z., Shen, Y., Wang, P., Song, B., He, Y., . . . Chen, Y. (2019). Organic frameworks induce synthesis and growth mechanism of well-ordered dumbbell-shaped ZnO particles. *Materials Chemistry and Physics, 232*, 129–136.

Issa, B., Obaidat, I. M., Albiss, B. A., & Haik, Y. (2013). Magnetic nanoparticles: surface effects and properties related to biomedicine applications. *International Journal of Molecular Sciences, 14*(11), 21266–21305.

Jing, Z., Shi, X., Zhang, G., & Qin, J. (2015). Synthesis, stereocomplex crystallization and properties of poly (l-lactide)/four-armed star poly (d-lactide) functionalized carbon nanotubes nanocomposites. *Polymers for Advanced Technologies, 26*(3), 223–233.

Khan, I., Saeed, K., & Khan, I. (2019). Nanoparticles: properties, applications and toxicities. *Arabian Journal of Chemistry, 12*(7), 908–931.

Kolhatkar, A. G., Jamison, A. C., Litvinov, D., Willson, R. C., & Lee, T. R. (2013). Tuning the magnetic properties of nanoparticles. *International Journal of Molecular Sciences, 14*(8), 15977–16009.

Kristensen, D. B., Imamura, K., Miyamoto, Y., & Yoshizato, K. (2000). Mass spectrometric approaches for the characterization of proteins on a hybrid quadrupole time-of-flight (Q-TOF) mass spectrometer. *ELECTROPHORESIS: An International Journal, 21*(2), 430–439.

Kumar, A. P., Depan, D., Tomer, N. S., & Singh, R. P. (2009). Nanoscale particles for polymer degradation and stabilization—trends and future perspectives. *Progress in Polymer Science, 34*(6), 479–515.

Langevin, D., Lozano, O., Salvati, A., Kestens, V., Monopoli, M., Raspaud, E., Driessen, M. (2018). Inter-laboratory comparison of nanoparticle size measurements using dynamic light scattering and differential centrifugal sedimentation. *NanoImpact, 10*, 97–107.

Li, J., & Zhang, J. Z. (2009). Optical properties and applications of hybrid semiconductor nanomaterials. *Coordination Chemistry Reviews, 253*(23–24), 3015–3041.

Li, N., Zhao, P., & Astruc, D. (2014). Anisotropic gold nanoparticles: synthesis, properties, applications, and toxicity. *Angewandte Chemie International Edition, 53*(7), 1756–1789.

Mayeen, A., Shaji, L. K., Nair, A. K., & Kalarikkal, N. (2018). Morphological characterization of nanomaterials *Characterization of Nanomaterials* (pp. 335–364): Elsevier.

Mitrano, D., Ranville, J. F., Bednar, A., Kazor, K., Hering, A. S., & Higgins, C. P. (2014). Tracking dissolution of silver nanoparticles at environmentally relevant concentrations in laboratory, natural, and processed waters using single particle ICP-MS (spICP-MS). *Environmental Science: Nano, 1*(3), 248–259.

Mørup, S., Madsen, D. E., Frandsen, C., Bahl, C. R., & Hansen, M. F. (2007). Experimental and theoretical studies of nanoparticles of antiferromagnetic materials. *Journal of Physics: Condensed Matter, 19*(21), 213202.

Mourdikoudis, S., Pallares, R. M., & Thanh, N. T. (2018). Characterization techniques for nanoparticles: comparison and complementarity upon studying nanoparticle properties. *Nanoscale, 10*(27), 12871–12934.

Mukhopadhyay, S. M. (2003). Sample preparation for microscopic and spectroscopic characterization of solid surfaces and films. *Chemical Analysis-New York-Interscience Then John Wiley-*, 377–412.

Oktaviani, O. (2021). Nanoparticles: properties, applications and toxicities. *Journal Latihan, 1*(2), 11–20.

Ong, Q., Luo, Z., & Stellacci, F. (2017). Characterization of ligand shell for mixed-ligand coated gold nanoparticles. *Accounts of Chemical Research, 50*(8), 1911–1919.

Pal, S. L., Jana, U., Manna, P. K., Mohanta, G. P., & Manavalan, R. (2011). Nanoparticle: an overview of preparation and characterization. *Journal of Applied Pharmaceutical Science 1*(6), 228–234.

Paramasivam, G., Palem, V. V., Sundaram, T., Sundaram, V., Kishore, S. C., & Bellucci, S. (2021). Nanomaterials: synthesis and applications in theranostics. *Nanomaterials, 11*(12), 3228.

Praetorius, A., Gundlach-Graham, A., Goldberg, E., Fabienke, W., Navratilova, J., Gondikas, A., von der Kammer, F. (2017). Single-particle multi-element fingerprinting (spMEF) using inductively-coupled plasma time-of-flight mass spectrometry (ICP-TOFMS) to identify engineered nanoparticles against the elevated natural background in soils. *Environmental Science: Nano, 4*(2), 307–314.

Sang, X., Lupini, A. R., Ding, J., Kalinin, S. V., Jesse, S., & Unocic, R. R. (2017). Precision controlled atomic resolution scanning transmission electron microscopy using spiral scan pathways. *Scientific Reports, 7*(1), 1–12.

Sangiorgi, N., Aversa, L., Tatti, R., Verucchi, R., & Sanson, A. (2017). Spectrophotometric method for optical band gap and electronic transitions determination of semiconductor materials. *Optical Materials, 64*, 18–25.

Sanjay, S. S., & Pandey, A. C. (2017). A brief manifestation of nano-technology *EMR/ESR/EPR spectroscopy for characterization of nanomaterials* (pp. 47–63): Springer.

Sayes, C. M., & Warheit, D. B. (2009). Characterization of nanomaterials for toxicity assessment. *Wiley Interdisciplinary Reviews: Nanomedicine and Nanobiotechnology, 1*(6), 660–670.

Sheikhzadeh, E., Beni, V., & Zourob, M. (2021). Nanomaterial application in bio/sensors for the detection of infectious diseases. *Talanta, 230*, 122026.

Shin, S. W., Yuk, J. S., Chun, S. H., Lim, Y. T., & Um, S. H. (2020). Hybrid material of structural DNA with inorganic compound: synthesis, applications, and perspective. *Nano Convergence, 7*(1), 1–12.

Singh, M. P., & Strouse, G. F. (2010). Involvement of the LSPR spectral overlap for energy transfer between a dye and Au nanoparticle. *Journal of the American Chemical Society, 132*(27), 9383–9391.

Tian, Y., & Wu, J. (2018). A comprehensive analysis of the BET area for nanoporous materials. *AIChE Journal, 64*(1), 286–293.

Wall, J., & Hainfeld, J. (1986). Mass mapping with the scanning transmission electron microscope. *Annual Review of Biophysics and Biophysical Chemistry, 15*(1), 355–376.

Wirth, R. (2009). Focused Ion Beam (FIB) combined with SEM and TEM: advanced analytical tools for studies of chemical composition, microstructure and crystal structure in geomaterials on a nanometre scale. *Chemical Geology, 261*(3–4), 217–229.

Wu, C., Zhou, X., & Wei, J. (2015). Localized surface plasmon resonance of silver nanotriangles synthesized by a versatile solution reaction. *Nanoscale Research Letters, 10*(1), 1–6.

Xu, Y., Jiang, S., Simmons, C. R., Narayanan, R. P., Zhang, F., Aziz, A.-M., Stephanopoulos, N. (2019). Tunable nanoscale cages from self-assembling DNA and protein building blocks. *ACS Nano, 13*(3), 3545–3554.

Yang, D. T., Lu, X., Fan, Y., & Murphy, R. M. (2014). Evaluation of nanoparticle tracking for characterization of fibrillar protein aggregates. *AIChE Journal, 60*(4), 1236–1244.

Yang, H., Coombs, N., & Ozin, G. A. (1997). Morphogenesis of shapes and surface patterns in mesoporous silica. *Nature, 386*(6626), 692–695.

# 4

# CHEMICAL CHARACTERIZATION
# TECHNIQUES IN DETAIL

## 4.1 OPTICAL SPECTROSCOPY

The ability to tune electrical and optical behavior, chemical reactivity, and mechanical or structural stability are a few of the special qualities that nanoscale materials possess. Nanomaterials are appealing for a wide range of applications including sensor technology, imaging, and catalysis because of their inherent characteristics (Chavali & Nikolova, 2019).

Nanomaterials have been studied using a variety of methods. Nanomaterials can interact with electromagnetic radiation (light) or electrons in the science of spectroscopy or microscopy, respectively. These interactions reveal important details regarding the shape, content, and crystal structure of nanomaterials, as well as their optical characteristics (Myroshnychenko et al., 2012). This section focuses on most important techniques and principles that aid in the chemical characterization of nanomaterials.

With regard to the interaction of matter and light, optical spectroscopy entails a wide range of electromagnetic waves from long wavelength, low-energy radio waves to short wavelength, high-energy gamma rays (Tkachenko, 2006). The light's wavelength, intensity, and impact on the molecules or atoms in the sample all have a role in how light interacts with the matter. Depending on the energy state transitions of the molecules, atoms, or ions in the sample, light can scatter, absorb, or emit when it strikes the sample. The sample's reflectance, transmittance, and absorption spectra cab be obtained from a spectrometer.

Photoluminescence (PL), Fourier transform infrared spectroscopy (FT-IR), Ultraviolet-visible-near-infrared (UV-Vis-NIR), and Raman spectroscopy are a some optical spectroscopy methods that can be used to describe nanomaterials (Boukhoubza et al., 2020). The study of electronic absorption, or UV-Vis-NIR spectroscopy

DOI: 10.1201/9781003354185-4

which encompasses the ultraviolet-visible-near-infrared spectrum exposes information about the electronic structure of a sample by showing the electronic excited states and phonon replicas in the resulting optical spectrum (Manuel & Shankar, 2021). As it can directly detect the spectral absorption behavior of plasmonic metal NPs, UV-Vis-NIR spectroscopy provides an easy way to optically characterize these nanomaterials. This assists in identifying the relevant wavelength that corresponds to the plasmon resonance (collective electronic charge oscillations in metal NPs that are excited by light and result in enhanced near-field absorption at the resonance wavelength). Utilizing UV-Vis-NIR spectra, linewidth measurements of plasmon resonance peaks reveal local differences in host particle permittivity, variability in NP size, coupling effects, and polydispersity as well as aggregation (Loiseau et al., 2019).

## 4.1.1 Optical absorption spectroscopy (OAS)

The majority of work in optical absorption spectroscopy (OAS) has focused on IR micro-spectroscopy but ordinary microscopes have been modified into portable or DIY spectrophotometers for a range of uses. It has access to the UV-visible region for identifying and/or mapping DNA and dyes in biology or forensics (Chassé et al., 2015). UV-Vis micro-spectroscopy is also used in chemistry and materials science but its spatial resolution, spectral range, and detection limits were more limited than those of standard laboratory spectrophotometers. In earth science, in situ high-temperature visible micro-spectroscopy has been developed to study the kinetics of color change with temperature in volcanic materials. In the special case of transition elements, electronic transitions span a wide energy range thus requiring an extended wavelength range to study the spectroscopic properties of microscopic samples. Recent research has shown interest in developing dedicated micro-spectrophotometers and adapting microscopes to laboratory spectrophotometers. For example, μ-OAS was recently developed in the reflection mode in the range 400–1,600 nm (Chassé et al., 2015). In the transmission mode, some examples of optical microscopes that were self-made into UV-Vis-NIR spectrophotometers have also been reported which measure the infrared and visible regions separately. The next step is to find a versatile microscope that can do OAS measurements that operate from UV to NIR without chromatic aberration and enable spatially resolved OAS in situations that may have fluctuating pressure and/or temperature. A versatile micro-spectrophotometric technique that enables routine measurements of optical absorption spectra over a wide range (200–3,300 nm) was reported by Chassé et al. (2015). This spectrophotometer-equipped optical microscope has totally reflecting optics in contrast to the conventional optical microscopes which have glass lens optics and display chromatic aberration over a wide range of wavelengths. The visible spectral range guarantees low signal loss coverage of the whole IR spectral range (50–10,000 cm$^{-1}$). The central element is a pair of Schwarzschild-type (NA = 0.54) reflective Cassegrain design condenser lenses that focus the light onto the sample and collect the light transmitted through the sample. It can record spatially resolved spectra with beam diameters from 20 to 120 μm and can be combined with a hot stage for investigations at various temperatures. This Cassegrain microscope is designed to fit in the sample chamber of a dual-beam UV-Visible-NIR spectrophotometer. The efficiency of this setup is demonstrated by two μ-OAS studies.

## 4.1.2 Photoluminescence (PL)

Photoluminescence spectroscopy, often called PL, is when light energy or photons stimulate the emission of photons from any material. It is a non-contact, non-destructive method of inspecting materials. Basically, light is absorbed and directed into the sample where a process called photoexcitation occurs. The material undergoes photoexcitation which allows it to go into a higher electronic state and then relax back to a lower energy level while releasing energy (photons). PL is the process through which light or luminescence is emitted. By combining Raman analysis and PL detection, we can characterize both the vibrational and electronic properties of materials on a single benchtop platform. Combining the Raman-PL system enables confocal mapping capabilities with sub-micron spatial resolution. A wide range of excitation wavelengths from UV to NIR is possible and the depth of penetration into the material can be controlled thus controlling the sample volume. PL used in fluorescence spectroscopy can produce two results: fluorescence and phosphorescence. The quantity of photons release is known as the photoluminescence quantum yield or PLQY of a molecule or other substance. This is due to the fact that one of the widely used methods of fluorescence spectroscopy is part of the number of absorbed photons. When light energy or photons induce the emission of photons, photoluminescence occurs. It has three forms of photo-luminescence which include fluorescence, phosphorescence, and chemiluminescence. The process by which photons stimulate molecules into an electrically excited state is known as fluorescence. The lowest singlet excited state releases photons after the excited state experiences a rapid thermal energy loss to the surroundings through vibration. Other non-radiative processes like energy transfer and heat loss compete with this photon emission process. Spectrometers and microscopes can be used for time-resolved photoluminescence (TRPL) (Metzger et al., 2005). Instruments with picosecond lasers using time-correlated single photon counting (TCSPC) are used as excitation sources in the solar industry to monitor carrier lifetime and cell performance (O'Connor, 2012).

TRPL is a technique that allows measuring changes in the PL intensity of a sample over time in response to a laser pulse. The equipment used is similar to that of a standard PL but with some differences. Instead of a spectrometer, a monochromator is used which only picks one wavelength of light to be fed into a photomultiplier tube detector that can count individual photons. The wavelengths chosen correspond to PL emissions from samples identified using conventional PL measurements. The computer's charge coupled device (CCD) input is also connected to a detector near the output of the laser source that is used to trigger the measurement. TRPL uses ~150 fs duration laser pulses fired at MHz frequency. A pulse that leaves from the laser acts as a trigger to start measurements at the CCD input. Upon hitting the sample, the pulse induces a PL emission that decreases in intensity over time before being hit by the next pulse. A computer may plot a light intensity decay curve for light of a specific wavelength by counting the photons that hit the detector over brief time periods. To increase the signal-to-noise ratio, the curves of succeeding pulses are also added. The decay time of the respective sample's PL can provide information about the processes running in it. An example of an experimentally obtained PL decay curve is shown in Figure 4.1. A laser pulse initiates a rapid increase in PL intensity that decays more slowly over a few nanoseconds. PL is a major contactless optical technique used to assess the crystalline clarity and purity of energy device materials as well as to pinpoint some contaminants.

A typical example of PL spectra is shown in Figure 4.1. The $Zn_xCd_{1-x}$ S NPs stabilized by carboxymethylcellulose sodium salt are shown in their PL spectra. The $Zn_xCd_{1-x}S$ quantum dots had

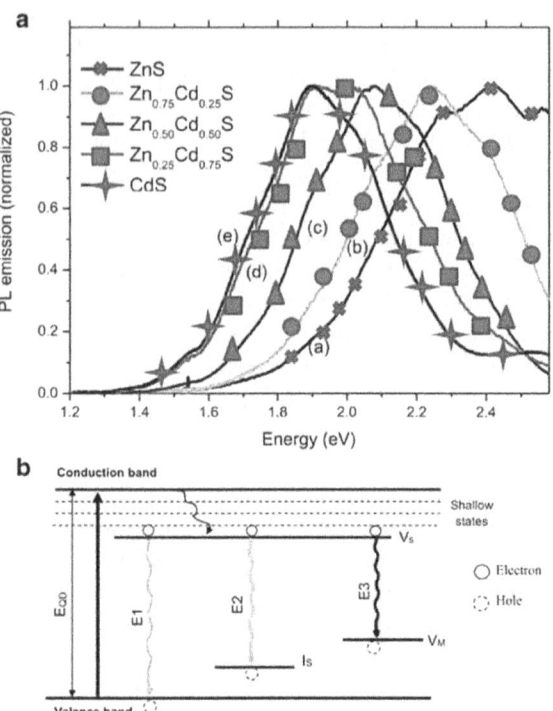

**Figure 4.1** Photoluminescence spectroscopy of conjugates. (a) PL emission spectra for $Zn_x Cd_{1-x} S$ ($x$ = 1.0 (a), 0.75 (b), 0.50 (c), 0.25 (d), and 0 (e). (b) The energy band structure suggested for three main emissions of alloyed $Zn_x Cd_{1-x} S$ quantum dots (Mansur, Mansur, Caires, Mansur, & Oliveira, 2017).

multicolored light emissions that were mostly caused by the photoluminescence that was defect activated. This means that the lattice point defects such as vacancies and interstitial atoms effectively trap electrons, holes, and exciton charge carriers at energies below the band-to-band optical transition. The emission path of the electron-hole pairs produced by irradiation depends on the presence of the traps, its density, and respective depths. According to the synthesis method followed in this study, vacancies of metals ($V_M$) such as $V_{Cd}$ or $V_{Zn}$ and the presence of sulfur in the interstitial sites of the lattice ($I_S$) are expected when using the stoichiometric molar ratio of $M^{2+}:S^{2-}$ = 1:2. Additionally, the creation of sulfur vacancies ($V_s$) at the quantum dot surface may result from the strong interaction between metallic cations and the anionic carboxylate species from carboxymethylcellulose which are entirely deprotonated at the pH of the synthesis. In this regard, considering the combination of the types of point defects ($V_M$, $V_{Cd}$, $V_{Zn}$, and $I_s$) and the similarity of the emission curves for $Zn_x Cd_{1-x} S$, three main radiative emissions were observed in the NPs which were assigned to the energy band diagram depicted in Figure 4.1b (Mansur et al., 2017).

## 4.1.3 Fourier transform infrared spectroscopy (FTIR)

FTIR spectroscopy is a specific type of IR spectroscopy (Riaz et al., 2018). Unlike a dispersive IR spectrometer where a dispersive element splits the incoming light into its spectral components and where each component is measured individually, one at a time ("scanned"), in the FTIR, all frequencies of light are measured simultaneously. The IR spectrum is then obtained via a mathematic conversion called Fourier transformation. Since FTIR spectroscopy measures all frequencies simultaneously, FTIR analysis can be performed faster compared to a scanning technique.

In FTIR spectroscopy, all spectral components of the light source are detected together. To obtain an FTIR spectrum, the spectral composition of the incoming light is permanently modified. Therefore, the signal detected by the FTIR detector is time-dependent. A mathematical operation called Fourier transformation allows the conversion of the detector signal from a time domain into the frequency domain (ultimately displayed in wavenumbers, $cm^{-1}$), which results in the well-known FTIR spectrum. FTIR spectrometers require three basic components: an IR light source that emits the IR light, an interferometer that modifies the spectral composition of the IR light in a time-dependent manner and a detector that detects the light intensity.

When the IR light enters the interferometer, a beam splitter divides the light into two optical beams. The first beam is reflected by a fixed mirror while the second beam is reflected by a moving mirror. The latter mirror constantly moves back and forth and depending on its position, the second beam travels either a longer or shorter distance. The two beams meet again at the beam-splitter where they interfere with each other. Due to the variation of the traveling distance of the second beam, the resulting IR light exiting the interferometer has a constantly changing frequency distribution. The detector records this "interferogram"; a function of the signal intensity versus time (= mirror position) which is Fourier transformed by a computer into a frequency spectrum; a function of the signal intensity versus frequency/wavenumber.

There is no difference between IR and FTIR. FTIR is a more specific term that describes an IR measurement in which the FTIR principle has been applied (recording of the signal in the time domain and subsequent Fourier transformation of the signal into the frequency domain). FTIR spectroscopy is the state-of-art technique for performing IR spectroscopy. Hence, as with all other IR spectroscopy techniques, FTIR spectroscopy measures the energy required to initiate molecular vibrations in a sample. One of the great advantages of FTIR spectroscopy is that it provides both qualitative and quantitative information from the same recorded FTIR spectrum.

FTIR spectroscopy is a very robust and reliable technique for the positive identification of unknown samples. The collected IR spectrum of an unknown sample is compared to a great number of IR spectra of known compounds. Each compound features a rather unique IR spectrum and if the measured IR spectrum of an unknown sample matches one of the known spectra, it is fair to say that these compounds are identical. Normally, the spectral matching is automatically done using suitable search algorithms and FTIR libraries (Lowry, Huppler, & Anderson, 1985). Libraries can be simply created with a few clicks from collected spectra, or commercially available libraries can be used.

It is also possible to extract quantitative information from an IR spectrum. For example, this could be the position of a certain IR band. However, in most cases, it is used to determine the concentration of a substance in a sample. The IR spectra of standard samples with known concentration can be used to create a correlation curve (French, Simic, & Thevenin, 2020). This correlation curve is then used to determine the concentration of the unknown sample from its IR spectrum. These "quantification models" that are based on signal intensity, band area, or chemometric approaches are also used for more challenging correlations (Weigel, Gehrke, Recknagel, & Stephan, 2021). Due to its many advantages over other technologies and its flexibility, FTIR analysis has become an established analytical technique in numerous fields of application and industries from quality control to research. FTIR testing of incoming goods ensures that the correct raw material is delivered and only the raw material that meets quality standards enters the production process. Equally the important factor is the quality control of the finished products by FTIR spectroscopy. Further uses of routine FTIR analysis are the identification of counterfeits, the detection of contaminants, and the troubleshooting of production problems. FTIR spectroscopy has some intrinsic advantages. A FTIR spectrometer measures all frequencies/wavenumbers simultaneously. Therefore, FTIR spectroscopy is a fast and easy analytical technique that provides answers within seconds. FTIR analysis is also considered resource friendly as it only requires a small amount of sample, little or no sample preparation and no consumables. FTIR spectroscopy is also flexible and versatile. It can be applied to a variety of sample types including liquids, solids, powders, semisolids, gases, and pastes, and it delivers quantitative as well as qualitative information. The two most important sampling techniques are transmission FTIR and attenuated total reflectance FTIR (ATR-FTIR). The other important FTIR sampling techniques include spectral reflection and diffuse reflection:

### 4.1.3.1 Spectral reflection – liquids, thins films, bulk materials

Spectral reflection sampling can be used for FTIR testing of liquids, thin films, and bulk materials. The IR light hits the sample at a defined angle and is reflected by the surface (Khoshhesab, 2012). The reflected light is detected at the same defined angle as the incoming light.

### 4.1.3.2 Diffuse reflection – mainly powders

The sampling technique of diffuse reflection is slightly different to the aforementioned technique. Diffuse reflection is mainly used for the FTIR analysis of powders of nanomaterials. Here, the IR light hits the sample and is reflected in all directions.

The traditional way of performing FTIR spectroscopy is in transmission mode. Transmission FTIR spectroscopy is suitable for liquids, gases, powders, and films. In transmission spectroscopy, the IR light that passes through the sample is detected by the FTIR detector on the other side of the sample. The FTIR analysis of liquid samples using demountable or flow cells is particularly challenging (Nasse,

Ratti, Giordano, & Hirschmugl, 2009). The use of such cells can be complex, error-prone and time-consuming for the following reasons:

- Cells are fragile and complex to assemble
- The design makes it difficult to get a reproducible path length
- Cells tend to leak
- Air bubbles can interfere with the analysis
- Cleaning and assembling of cells can be lengthy procedures
- Sticky and viscous samples are hard to introduce
- Significant amount of sample volume and rinsing solvent are required

ATR-FTIR gained its popularity due to the ability to quickly and easily measure a broad range of sample types including liquids, solids, powders, semisolids, and pastes. In ATR sampling mode, the IR light that travels through a crystal is internally reflected at the crystal-sample interface and the reflected light then travels to the FTIR detector. During the internal reflection, a part of the IR light travels into the sample where it can be absorbed. The portion of light that enters the sample is called the evanescent wave. The evanescent wave's penetration depth into the sample is determined by the reflective index difference between the ATR crystal and the sample. Therefore, different crystal materials are used in ATR-FTIR analysis depending on the sample type. Diamond is the most widely applicable material but zinc selenide or germanium ATR crystals are also used for certain applications (Khoshhesab, 2012).

## 4.1.4 Raman spectroscopy

Raman scattering is the technique that involves stimulated molecules at higher energy levels scattering light. The Raman effect is an alternative term for the aforementioned effect. The stimulating monochromatic beam needs to be extremely intense for the molecule to enter a virtual energy state (laser beam). Most molecules relax back to their initial states of S0, where light with the same wavelength as the stimulating light is released (Rayleigh scattering) (Vandenabeele, 2013). The emitted photons have less energy than the excited photons because only a very tiny portion of the excited molecules relax back to a vibrationally stimulated state (Stokes lines) (Szymanski, 2012). Raman scattering is always of very low intensity, and because very few molecules adopt this relaxation pathway, its research needs high-quality equipment. After being excited, molecules that were previously in a vibrationally excited state can relax to the ground state, generating photons with a higher energy than those that originally excited them (anti-Stokes lines) (Hof, 2003).

The $\exp\left(-\dfrac{E}{bT}\right)$, where E is the energy difference between the ground and excited states, b is the Boltzmann's constant, and T is the temperature, gives the percentage of molecules that are initially in the vibrationally excited state. As a result, the Anti-Stokes lines have significantly lower intensities than the Stokes lines and using the ratio of their intensities, it allows for in situ temperature monitoring. The condition for a molecule to be Raman active is a change in the polarization (deformation) of the electron cloud during the interaction with the incident radiation (Hof, 2003).

$$\mu^` = \varkappa_1 E + \frac{1}{2}\varkappa_2 E^2 + \frac{1}{6}\varkappa_3 E^3 \ldots\ldots$$

The dipole moment is a two-dimensional (vector) term, whereas $\varkappa_i$ is the molecular polarizability which is a three-dimensional (tensor) term, and may be simplified to its linear term at commonly used field strength levels (laser power up to 1 kW per line) is applicable. Only when the exciting light is extremely intense do nonlinear terms need to be considered (above 1 MW per line). Due to this circumstance, the conventional Raman effect is frequently referred to as a "linear Raman effect" as opposed to "nonlinear Raman effects" which are visible with very powerful laser stimulation (Hof, 2003). $\varkappa_2$ contributions result in the hyper-Raman effect, $\varkappa_3$ contributions result in stimulated Raman scattering (SRS) and coherent anti-Stokes Raman spectroscopy (CARS) (Duboisset et al., 2015).

Two photons are used to stimulate the SRS process. The only Raman modes that produce stimulated Stokes emissions are those with the highest gain factors. If the molecular system is given a strong external Stokes field by utilizing a second laser beam for excitation, all active Raman modes can be seen theoretically (three wave mixing) (Shipp, Sinjab, & Notingher, 2017). When two electromagnetic waves with frequencies $\dot{\omega}1$ and $\dot{\omega}2$ are superimposed, a third electromagnetic wave is produced at the difference frequency (difference frequency generation, DFG, $\dot{\omega}1-\dot{\omega}2$), second harmonic (second harmonic generation, SHG, $\dot{\omega}1 + \dot{\omega}2$) or sum frequency (sum frequency generation, SFG, $\dot{\omega}1 + \dot{\omega}2$) depending on the value of $\dot{\omega}i$. The square of the vibrational amplitudes determines the harmonic intensities that the oscillating dipoles emit (Hof, 2003).

The Raman method is an alternative to IR spectroscopy, which approaches the excited vibrational state indirectly (Wright, 1980). The plot of Raman intensity versus Raman shift is known as the Raman spectrum. The Raman shift, band intensity and band shape are the three parameters of a Raman band (Cancado et al., 2008). Similar to the IR spectrum, a nanomaterial's chemical structure directly affects the characteristics of a Raman spectrum such as the number of Raman bands, their shapes and their intensities. The differing excitation conditions; a change in the dipole moment (a vector variable) for the IR spectrum and a change in the polarization (a tensor quantity) for the Raman spectrum are the basis for the complementarity of the two spectra.

In general, the quantitative analysis methods for single and multicomponent analysis as well as the qualitative analysis by group frequencies are the same as in IR. Raman's use is severely constrained by the fluorescence phenomenon. Raman scattering can be up to 107 times weaker than fluorescence (Hof, 2003). Therefore, it may be impossible to observe the analyte's Raman spectra due to the fluorescence of minute contaminants. The gap between the virtual energy state and the electrically excited state S1 must be sufficiently broad (excitation wavelength selection amongst UV, MIR, and NIR) in order to prevent the masking of Raman scattering by fluorescence (de Oliveira Penido, Pacheco, Lednev, & Silveira Jr, 2016).

## 4.2 ELECTRON SPECTROSCOPY

### 4.2.1 Energy dispersive X-ray spectroscopy (EDS)

The analytical technique of energy dispersive X-ray spectroscopy (EDS, also known as EDX or XEDS) allows for the chemical characterization and elemental analysis of materials (Mishra, Zachariah, & Thomas, 2017). By emitting core-shell electrons, a sample energized by an energy source (such an electron beam from an electron microscope) releases some of the absorbed energy. The respective area is subsequently filled with higher-energy outer shell electrons which emit X-rays due to an energy difference with a distinct spectrum dependent on the initial atom (Shirley & Jarochowska, 2022). This makes it possible to determine the makeup of a sample volume that has been excited by an energy source. The element is identified by its position on the spectral peak and the concentration of the element is indicated by the signal's intensity. As was already established, the electron beam has sufficient energy to produce an X-ray emission as well as core-shell electron emission. The compositional data down to the atomic level can be obtained by integrating an EDS detector into the electron microscope (Wirth, 2009). Characteristic X-rays are released from the sample and detected as the electron probe passes over it. A particular location in the sample is allocated to each recorded EDS spectra. Signal strength and spectral purity affect the quality of the results. The signal strength is heavily dependent on a good signal-to-noise ratio (allowing faster recordings and artifact-free results) for trace element identification and dosage minimization in particular (Ernst et al., 2014). The quantity of obvious flaws is influenced by cleanliness. This is a result of the electron column's constituent materials. An example of EDX spectra is shown in Figure 4.2a.

### 4.2.1.1 EDS elemental analysis

Fundamental uses for EDS include elemental analysis the images from electron microscopy are enhanced with compositional data from EDS, giving a comprehensive perspective on the morphological and chemical nature of the sample. Enhancing the speed of the technique and sensitivity will be crucial as EDS analysis merges more and more with standard electron microscopy (Lapresta-Fernández et al., 2014).

### 4.2.1.2 EDX elemental mapping

Efficient EDS detection systems enable rapid compositional analysis with sub-nanometer resolution. Atomic-resolution spectroscopy is now commonly possible with electron microscopy owing to the atomic size electron probes and due to the X-ray detectors of high current and high sensitivity. Due to its distinctive X-ray signal, EDS, is an excellent method for identifying individual atoms. Researchers can characterize and alter materials at the atomic level using this knowledge which provides previously unattainable insight into the behavior of nanomaterials and particles. Figure 4.2 demonstrates how distinct chemical signals can be used to identify individual atomic locations of a respective sample. Apart from being visible, individual atom columns also stand out from surrounding columns because of their great contrast. Additionally, the EDS signal makes it possible to detect light elements which are notoriously challenging to be detected at these resolutions (Susi & Byler, 1986). The low probe current necessary for atomic-scale spatial resolution is no longer a barrier to the acquisition and interpretation of X-ray spectra thanks to advancements in signal creation and detection. To determine a material's true elemental distribution or composition, compositional analysis must be performed. The resulting 3D chemical maps then reveal additional information about the structure-function correlations of a sample. Nanoscale analysis in 3D is becoming more and more important in contemporary materials research (Weyland, Yates, Dunin-Borkowski, Laffont, & Midgley, 2006). 3D EDS is a crucial technique since a complete 3D characterization contains both chemical and image data. Therefore, for the best possible outcomes, devices with dynamic, high-resolution imaging capabilities and quick, quantitative data capture are needed. True 3D structural and compositional study of nanomaterials is possible thanks to the versatility of acquisition techniques (TEM, STEM, and EDS), the simplicity and reproducibility of experiment optimization, and the quick and sensitive capture of elemental distribution data. By gradually changing the sample's viewing angle, electron tomography creates a 3D reconstruction of the substance. To reflect the original sample volume, this results in a tilted series of images that may be digitally back projected (Genc et al., 2013). The combination of EDS spectra with electron microscopy (EM) pictures can yield extensive elemental information.

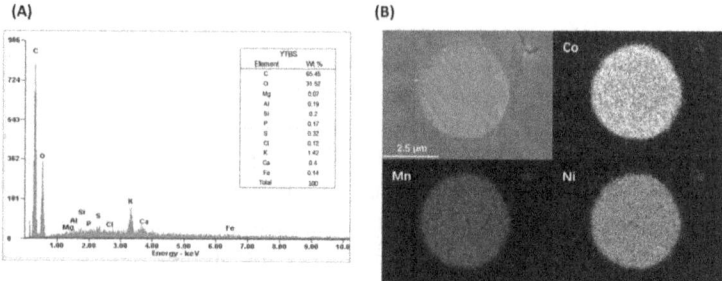

**Figure 4.2** (a) Energy dispersive X-ray analysis (EDX) spectrum of white yam (*Dioscorea rotundata*) tuber peel YTBS residue adsorbent (Asuquo, Martin, & Nzerem, 2018) and (b) SEM-EDX mapping photographs for Ni, Co, and Mn in products after hydrothermal treatment (Tang, Lu, & Luo, 2016).

## 4.2.2 X-ray photoelectron spectroscopy (XPS)

A surface sensitive analytical method known as XPS bombards the surface of a material with X-rays and then measures the kinetic energy of the emitted electrons. The surface sensitivity of this technique and the ability to extract information about the chemical states of the elements in the sample are two of the key qualities that make it an effective analytical method. Almost all materials, including semiconductors, polymers, textiles and dirt have their surfaces studied using XPS which can identify all elements except hydrogen and helium. The surfaces of each substance are what interact with the other materials. Surface wetting, adhesion, corrosion, charge transfer, and catalysis are just a few of the processes affected by the material surface and surface contaminants. Therefore, it is important to study and understand surfaces. The photoelectric effect, first identified by Heinrich Hertz in 1887, is the basis for XPS. He discovered that when surfaces are exposed to light, electrons are released. Albert Einstein formalized this idea in 1905, and for this effort he was awarded the Nobel Prize in Physics in 1921. Robinson and Rawlinson made the first observation of optical emission by irradiating X-rays in 1914, and Steinhardt and Serfass published the first use of X-ray emission as an analytical technique in 1951 (Hofmann, 2012). However, Kai Siegbahn at Uppsala University, Sweden in the 1950s and 1960s did most of the work to develop XPS into a method that we know today. For his work on high-resolution electron spectroscopy, commonly known as electron spectroscopy for chemical analysis (ESCA), he was awarded the Nobel Prize in 1981.

In XPS, the sample is irradiated with soft X-rays (energy less than ~6 keV) and the kinetic energy of the emitted electrons is analyzed (Figure 4.3). The emitted photoelectron is the result of the complete transfer of X-ray energy to a central electron. This is expressed as a mathematical equation as follows:

$$h\upsilon = BE + KE + \phi_{spec}$$

It simply exhibits that the energy of the X-ray is equal to the summation of the binding energy (BE) of the electron (how strongly it is bound to the atom/orbital), the kinetic energy (KE) of the emitted electron, and the spectrometer working function ($\Phi_{spec}$), a constant value (Stevie & Donley, 2020).

It is important to note that the photoelectron binding energy is measured relative to the Fermi level of the sample (not the vacuum level) which is why $\Phi_{spec}$ is included. The photoelectron spikes are represented by the element and the trajectory from which they are ejected. For example, "O 1s" describes electrons emitted from the 1s orbital of an oxygen atom. Any electron with a binding energy lower than that of the X-ray source must be emitted from the sample and observed using the XPS technique. The binding energy of an electron is a property of matter and is independent of the X-ray source used to emit it. When experiments are performed with different X-ray sources, the photoelectron binding energy will not change; however, the kinetic energy of the emitted photoelectrons will change as described in above equation (Fadley, 2010).

The loss of the central XPS electron results in a central "hole." This excited ionized state will expand by filling the hole with an electron from a valence orbital. This relaxation releases energy by one of two competing processes: X-ray fluorescence or Auger electron emission. X-ray fluorescence is not detected in the electron spectrum and will not be considered further here. Auger electrons generated will be detected and commonly used in XPS for qualitative analysis. Auger peak notation is traditionally based on the K, L, and M nomenclature for atomic orbitals. For example, Auger's main oxygen peak is denoted by KLL, indicating that the first electron ejected comes from a K orbital, the central hole-filling electron comes from an L orbital, and the final ejection of the Auger electron also comes from an L orbital. Sub-symbols are sometimes used to distinguish between specific L, M, and N orbitals (Stevie & Donley, 2020). The Auger process consists of three different electron transitions and the kinetic energy of the ejected Auger electron is described in the below equation.

$$KE_{Auger} = BE(K) - BE(L_1) - BE(L_3)$$

### 4.2.3 Auger electron spectroscopy (AES)

AES or Auger is a surface-sensitive analytical technique that utilizes a high-energy electron beam as an excitation source. Atoms that are excited by the electron beam can subsequently relax leading to the emission of "Auger" electrons. The kinetic energies of the emitted Auger electrons are characteristic of elements present within the top 3–10 nm of the sample. AES is very useful when investigating

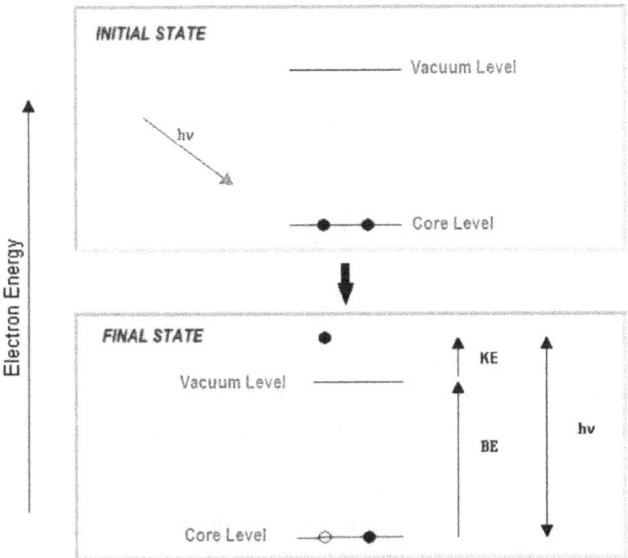

**Figure 4.3**  The photoemission process in X-ray photoelectron spectroscopy analysis. The solid circles represent electrons and the horizontal lines represent energy levels within the material being analyzed (Aziz & Ismail, 2017).

particles and small areas as it has the ability to investigate sizes smaller than 25 nm owing to the smaller diameter of the employed electron beam. It also provides a good alternative when thin films are too thin for EDS analysis. AES can sample thin film stacks to a depth of a micron or more with sputtering (Kuijers, Tieman, & Ponec, 1978). Above all, AES is a semi-quantitative method meaning that results are typically provided based on standard sensitivity factors provided by the equipment manufacturer. When more accurate results are required, these aforementioned factors can be obtained by looking at known compositions and comparing them to the unknown material (Sutton, Kriewall, Leu, & Newkirk, 2016).

The electron beam can be scanned over areas of variable sizes or it can be directly focused on a specific surface feature of interest. This ability to focus the electron beam to diameters of 10–20 nm makes AES an extremely useful tool for elemental analysis of small surface features. Other techniques that may also be considered are XPS and total reflection X-ray fluorescence TXRF (Chang, 1971).

When used in combination with a sputtering ion gun, AES can also perform compositional depth profiling.

### 4.2.4 Ultraviolet photoelectron spectroscopy (UPS)

Similar to XPS, UPS operates on a similar concept. The sole distinction is that, as opposed to >1 keV photons used in XPS, the photoelectric effect is induced using ionizing radiation with an energy of 10 eV (Béchu, Ralaiarisoa, Etcheberry, & Schulz, 2020). Gases like argon and neon can also be used to create ultraviolet photons in gas discharge lamps which are often filled with helium in the lab. Helium gas emits photons at energies of 21.2 eV (He I) and 40.8 eV respectively (He II). Since most nuclear level photoemissions are inaccessible using UPS due to the use of low-energy photons, spectral detection is restricted to the valence band region. There are two types of experiments performed using UPS which include the valence band detection and electron work function measurements. The valence band photoelectron signal originates from many molecular orbitals that have a high degree of hybridization, therefore, the peak binding energy shifts are much more variable and subtler than those seen for core level photoemission peaks. The majority of material characterization using valence band spectra involves spectral fingerprinting for this reason, and peak assignment is either done on surfaces with well-known electronic structures or in conjunction with computational research. These spectra are not used for quantification since the peak assignment of the valence band is ambiguous. Since the ionization cross-section of an orbital depends on the incident photon energy, different electronic transitions and states can be probed by using various photon energies, UPS is also frequently used to collect valence band spectra. When combined with XPS, this method can be very effective in studying the valence band. The short inelastic mean free path (IMFP) of free electrons within a solid is what gives XPS its inherent surface sensitivity. According to conventional wisdom, the "information depth" from which > 99% of a photoemission signal originates is defined as being at 3 mean free path lengths from the surface which in XPS is frequently quoted as 10 nm (Pintori & Cattaruzza, 2021). In comparison to XPS, UPS shows higher surface sensitivity. This is only a preliminary estimate because an electron's IMFP depends on both the physical characteristics of the solid medium it travels through and its kinetic energy, with lower kinetic energy electrons having shorter route lengths. With a roughly 2–3 nm information depth, UV photoemission spectroscopy uses lower incident photon energy to create photoelectrons with substantially lower kinetic energies than those detected by XPS (Brundle, 2002).

The difference between the Fermi level and the vacuum level is known as the electronic work function, a material property, which applies in the development of electronic devices. As a surface property, work function is strongly influenced by changes in composition or structure at the surface due to reasons such as atmospheric pollution. The electronic work functions were obtained spectroscopically by measuring the difference between the Fermi level and the "tail" cut off at the low kinetic end of the spectrum (i.e., the width of the spectrum) and by subtracting this value from the energy of the incident photons. This value can of course be measured with incident X-rays; however, UPS allows the calculation of the working function from a single spectrum. When the work function of electrons is measured by photoelectron spectroscopy, it is necessary to apply a small deviation (usually 5–10 V) to the sample surface, in order to separate the actual work function of the surface from that of the sample (Eland, 2013).

## 4.3 IONIC SCATTERING SPECTROSCOPY

Ion scattering spectroscopy (ISS) is a technique in which an ion beam is scattered by a surface. The kinetic energy of the dispersed ions is measured; observed peaks correspond to the elastic diffusion of atomic ions on the surface of the sample. Each particle on the surface of the sample produces a peak at a different measured kinetic energy brought about by the momentum movement between the incident ion and the atom. Scattered ions and scattered atoms usually have different masses but the total momentum of the atom and ion is conserved. Thus, when the initial "stationary" atom recoils, some kinetic energy is lost from the dispersed ion and the amount of energy lost depends on the relative masses of the atom and the ion. Ion scattering spectra typically contain a peak for each element in the sample, the separation of which is related to the relative atomic masses of those elements. In some cases, different isotopes of the same element can be resolved although this usually requires a primary ion heavier than $He^+$ (Krauss, Auciello, & Schultz, 1995). The strongly scattered signal is confined to the top atomic layer so the

ISS is extremely sensitive at the surface. Therefore, the sample must be clean and even small amounts of surface contaminants can significantly affect the ion scattering spectrum. The absolute kinetic energy ($E_S$) of each peak also depends on the energy of the scattered ion beam ($E_0$) because ISS is an energy loss technique. The quantity involved when handling the ISS is the ratio of the energies of the scattered and incident ions, $E/E_0$. The ion scattering angle ($\theta$) is also important. When the ion source and detector are fixed, the ion scattering angle is a parameter that must be taken into account (Baun, 1977).

$$\frac{E_S}{E_0} = Cos\theta \pm [[M_2 / M_1]2 - Sin^2\theta]2 / \left[1 + \frac{M\,2}{M\,1}\right]$$

The equation shows how the energy of a scattered ion peak is related to the other relevant factors.

$E_S$ – Kinetic energy of the scattered ion
$M_1$ – Relative atomic mass of the scattered ion
$E_0$ – Kinetic energy of the primary ion beam
$M_2$ – Relative atomic mass of the scattering surface atom
$\theta$ – Scattering angle

For a given device, $\theta$ is usually a constant since the angle between the ion gun and the input lens is 50°. $M_1$ is a constant for a given source gas, usually He, but sometimes Ne, Ar, and other inert gases are also used. $E_0$ is usually a constant in a given experiment. Therefore, this equation can be used to determine the mass of a scattering atom, $M_2$, from its peak position in the spectrum, or to predict the position of a peak with respect to a given atom (Nastasi, Mayer, & Wang, 2014). Sometimes $E_0$ is not known at the beginning of the test. The solution is to calibrate the energy using a standard, preferably, a pure known metallic element $M_2$ such as gold. Once the dispersion peak energy of the reference sample is measured, the only remaining unknown value, $E_0$, can be calculated. Once $E_0$ is known, the main goal of the ISS is to associate the ES dispersion energy peaks with atoms of mass $M_2$.

### 4.3.1 Ion scattering spectroscopy

Consider the example of ion scattering spectroscopy from Cu, Ag, and Au using a He+ ion with energy of 970 eV and a scattering angle of 130°. All specimens are lightly etched to remove most carbon and oxygen contaminants. The ISS spectra of copper and silver show some presence of residual oxygen at ~400 eV KE (Grant & Lambert, 1984). The Ag spectrum has some structure at ~60 eV. The low KE region of the ISS corresponds to the dispersed Ag ions producing a low-energy, high-intense kinetic energy spectrum from 0 to 200 eV KE.

## 4.3.2 Basic ions in ion scattering spectroscopy

Ion scattering spectra of a phosphorus-copper sample show the presence of oxygen, copper, and tin. This spectrum was acquired using He ions at 1 keV. Noble gases are commonly used for primary ions in the ISS which avoided possible surface contamination if more reactive materials are being used. Only surface atoms that can be detected are those with a mass greater than the mass of the primary ions (Taglauer, 1991). Therefore, helium provides the widest mass range and hydrogen is the only element that cannot be detected. Since the mass dissociation decreases with atomic mass, there is a value of about 400 eV for energy dissociation between O and Cu but between Cu and Sn, it is less than 100 eV. This result is obtained although the difference in mass between Sn and Cu is much larger than that between Cu and O. If a better mass resolution is desired, it is necessary to use noble gas ions as heavy as possible while being as light as possible.

## 4.3.3 Mass resolution by ion scattering spectroscopy

Helium is an ion for the ISS that allows elemental composition determination; however, mass unit resolution is lost for all target atoms heavier than Na due to the physics of the scattering process. The inherent full width at half maximum (FWHM) of peaks near $E_0$ is usually 2% of energy. In general, the mass resolution decreases rapidly as the mass increases. The mass resolution also depends on the angular range from which the ions are collected. Although the initial phosphorus-copper spectrum is obtained with an acquisition angle of 20°, this must be reduced to about 3° to obtain the second spectrum. The energy scattering in the ion beam will also affect the mass resolution as well as the degree of collimation of the beam (Young, Hoflund, & Miller, 1990).

## 4.4 RUTHERFORD BACKSCATTERING SPECTROMETRY (RBS)

RBS is a popular nuclear technique for studying the near surface layers of materials. Ions are fired at a target with energies in the meV range (usually 0.5–4 meV) and an energy-sensitive detector, often a solid state detector, captures the energy of the backscattered projectiles. RBS enables the quantitative analysis of a material's composition and the depth profiling of individual constituents. RBS has a very good sensitivity for heavy elements of the order of parts per million. Furthermore, this technique is quantitative without the need for reference samples, is non-destructive and has a strong depth resolution of the order of several nm (ppm) (Poehlman, 1980). For incident He ions, the examined depth is normally 2 m while it is 20 m for incident protons.

## 4.5 SECONDARY ION MASS SPECTROMETRY (SIMS)

A portion of the particles produced from the target is ionized when primary ions with energy levels of only a few keV sputter onto a solid sample. The process of examining these secondary ions using a mass spectrometer is known as secondary ion mass spectrometry. The elemental, isotopic, and molecular makeup of a solid surface exposed to ions can be learned from the secondary ion emission by that surface. The chemical environment and the sputtering circumstances will have a significant impact on the secondary ion yields (ion, energy, angle) (McPhail, 2006). As a result, the quantitative component of the technique may become more difficult. The most sensitive method for elemental and isotopic surface analysis is still SIMS. The primary beam focuses on the sample surface to generate ions, which are subsequently transformed into secondary ions in a mass spectrometer via a strong electrostatic potential. A high speed beam of neutral atoms (e.g., Ar) can be used to replace a beam of primary ions in a comparable technique, which is mostly utilized for surface investigation of compounds. Organic matter has few geosciences uses and is not treated here. In a vacuum, the interaction of the primary ion beam with the sample produces enough energy to ionize numerous elements. If the primary beam is made up of positively charged particles, the following ionization will favor the formation of negative ions, whereas the main negative ion beam will favor the formation of positive ions.

The primary ion beam surpasses the "static limit" in the "dynamic SIMS" mode, resulting in a large secondary ion yield. This approach is used to analyze a large number of elements and isotopes, and it is especially useful for analyzing isotopes and trace elements in minerals (e.g., REE in garnet). Furthermore, static SIM employs a significantly lower energy primary ion beam (usually Ga or Cs). This approach is often used to investigate atomic monolayers on the surface of materials in order to gather information on the material's molecular structures (e.g., organic compounds).

Several SIMS models are currently in commercial production for geoscience applications. The majority of these devices have a source region that controls the intensity, energy, and direction of the primary beam (relative to the sample). The ions created by this process form a secondary beam, which is subsequently sent to a mass spectrometer in a continuous high vacuum. The majority of SIMS devices used for elemental and isotope analysis work by accelerating ions generated by the source down a potential gradient, typically 10 kV, and then transferring those ions into the mass spectrometer. The specifics of mass spectrometer setups vary depending on application, but they invariably use both magnetic and electrostatic analyzers, which are referred to as fields. Forward geometry is used when the electrostatic analyser (area) comes before the magnetic region. The electrostatic filter decreases the energy range of the secondary ions, allowing them to be separated into independent ion beams (depending on the charge-to-mass ratio) by passing them through a magnetic field (magnetic field). Several ion beams can be measured simultaneously in this setup. If the magnetic zone comes before the electrostatic region (inverse geometry), mass resolution improves at the expense of the ability to measure numerous ion beams at the same time. SIMS sources can be integrated with other types of mass spectrometers, such as quadrupole and time-of-flight analyzers. In geosciences, the latter arrangements have less applicability.

The large radius direct and inverse geometry instruments that have been designed to analyze trace elemental and isotopic compositions in individual minerals/nanomaetrials with spatial resolution down to around 10 microns are some of the major facts of geoscience SIMS. U-Th-Pb geoengineering of zircon and other minor minerals is one of these uses. Furthermore, dual focus and large radius instruments can measure the isotopic composition of low atomic number elements like O with comparable spatial resolution, if not less for critical components.

Smaller radius and dual focus devices enable very high sensitivity (low limit of detection) for trace element analysis, roughly ten times that of microchips employing electron beams. These methods can also be used to "map" how specific elements in a sample are distributed. Other types of mass spectrometers are used by some SIMS (e.g., time-of-flight, quadrupole). The following facts are relevant while assessing the benefits of SIMS.

SIMS analyses consume very few samples (they are virtually non-destructive), for example, a typical U-Th-Pb study consumes only a few cubic micrometers of sample (Limoncelli, 2012). SIMS can test samples with low concentrations (down to the ppb level) due to its high sensitivity. As a result, SIMS is widely used in the semiconductor industry to detect trace components in non-conductive substances and is used to assess the amount of trace elements in meteorites, interplanetary dust, and other small samples. High sensitivity also enables a comprehensive understanding of elemental and molecular abundances and isotope ratios. In most situations, on-site analysis avoids the requirement for sophisticated sample preparation, allowing minerals to be evaluated immediately as particles or in thin form. Because atoms and molecules are formed when the sample sputters, not all constituents of all substrates (matrix) can be quantitatively evaluated. For example, intractable isotropic disturbances (equivalent masses of ions) in zircon are a barrier that cannot be overcome by direct geometry multi-gather devices or instruments with high-resolution inverse geometry. SIMS instrumentation is typically costly and is utilized mostly for surface characterization, molecular analysis, and depth profiling.

## REFERENCES

Asuquo, E. D., Martin, A. D., & Nzerem, P. (2018). Evaluation of Cd (II) ion removal from aqueous solution by a low-cost adsorbent prepared from white yam (Dioscorea rotundata) waste using batch sorption. *ChemEngineering, 2*(3), 35.

Aziz, M., & Ismail, A. (2017). X-ray photoelectron spectroscopy (XPS) *Membrane Characterization* (pp. 81–93): Elsevier.

Baun, W. (1977). Fine features in ion scattering spectra (ISS). *Applications of Surface Science, 1*(1), 81–102.

Béchu, S., Ralaiarisoa, M., Etcheberry, A., & Schulz, P. (2020). Photoemission spectroscopy characterization of halide perovskites. *Advanced Energy Materials, 10*(26), 1904007.

Boukhoubza, I., Khenfouch, M., Achehboune, M., Leontie, L., Galca, A. C., Enculescu, M., Jorio, A. (2020). Graphene oxide concentration effect on the optoelectronic properties of ZnO/GO nanocomposites. *Nanomaterials, 10*(8), 1532.

Brundle, C. R. (2002). *Electron spectroscopy*: Mittal Publications.

Cancado, L., Takai, K., Enoki, T., Endo, M., Kim, Y., Mizusaki, H., Pimenta, M. (2008). Measuring the degree of stacking order in graphite by Raman spectroscopy. *Carbon, 46*(2), 272–275.

Chang, C. C. (1971). Auger electron spectroscopy. *Surface Science, 25*(1), 53–79.

Chassé, M., Lelong, G., Van Nijnatten, P., Schoofs, I., de Wolf, J., Galoisy, L., & Calas, G. (2015). Optical absorption microspectroscopy (μ-OAS) based on Schwarzschild-type Cassegrain optics. *Applied Spectroscopy, 69*(4), 457–463.

Chavali, M. S., & Nikolova, M. P. (2019). Metal oxide nanoparticles and their applications in nanotechnology. *SN Applied Sciences, 1*(6), 1–30.

de Oliveira Penido, C. A. F., Pacheco, M. T. T., Lednev, I. K., & Silveira Jr, L. (2016). Raman spectroscopy in forensic analysis: identification of cocaine and other illegal drugs of abuse. *Journal of Raman Spectroscopy, 47*(1), 28–38.

Duboisset, J., Berto, P., Gasecka, P., Bioud, F.-Z., Ferrand, P., Rigneault, H., & Brasselet, S. (2015). Molecular orientational order probed by coherent anti-Stokes Raman scattering (CARS) and stimulated Raman scattering (SRS) microscopy: a spectral comparative study. *The Journal of Physical Chemistry B, 119*(7), 3242–3249.

Eland, J. H. D. (2013). *Photoelectron spectroscopy: an introduction to ultraviolet photoelectron spectroscopy in the gas phase*: Elsevier.

Ernst, T., Berman, T., Buscaglia, J., Eckert-Lumsdon, T., Hanlon, C., Olsson, K., Valadez, M. (2014). Signal-to-noise ratios in forensic glass analysis by micro X-ray fluorescence spectrometry. *X-Ray Spectrometry, 43*(1), 13–21.

Fadley, C. S. (2010). X-ray photoelectron spectroscopy: progress and perspectives. *Journal of Electron Spectroscopy and Related Phenomena, 178*, 2–32.

French, R. M., Simic, V., & Thevenin, M. (2020). Peak correlation classifier (PCC) applied to FTIR spectra: a novel means of identifying toxic substances in mixtures. *IET Signal Processing, 14*(10), 737–744.

Genc, A., Kovarik, L., Gu, M., Cheng, H., Plachinda, P., Pullan, L., Wang, C. (2013). XEDS STEM tomography for 3D chemical characterization of nanoscale particles. *Ultramicroscopy, 131*, 24–32.

Grant, R., & Lambert, R. (1984). Basic studies of the oxygen surface chemistry of silver: chemisorbed atomic and molecular species on pure Ag (111). *Surface Science, 146*(1), 256–268.

Hof, M. (2003). Basics of optical spectroscopy. *Handbook of Spectroscopy, 1*, 39–47.

Hofmann, S. (2012). *Auger-and X-ray photoelectron spectroscopy in materials science: a user-oriented guide* (Vol. 49): Springer Science & Business Media.

Khoshhesab, Z. M. (2012). Reflectance IR spectroscopy. *Infrared Spectroscopy-Materials Science, Engineering and Technology, 11*, 233–244.

Krauss, A. R., Auciello, O., & Schultz, J. A. (1995). Time-of-flight, ion-beam surface analysis for in situ characterization of thin-film growth processes. *MRS Bulletin, 20*(5), 18–23.

Kuijers, F. J., Tieman, B. M., & Ponec, V. (1978). The surface composition of platinum-palladium alloys determined by auger electron spectroscopy. *Surface Science, 75*(4), 657–680.

Lapresta-Fernández, A., Salinas-Castillo, A., Anderson De La Llana, S., Costa-Fernández, J. M., Domínguez-Meister, S., Cecchini, R., Sánchez-López, J. C. (2014). A general perspective of the characterization and quantification of nanoparticles: imaging, spectroscopic, and separation techniques. *Critical Reviews in Solid State and Materials Sciences, 39*(6), 423–458.

Limoncelli, M. (2012). *Short-term erosion pattern in the Alps-Apennines belt constrained by downstream changes of zircons morphology and U-Pb ages from the Po drainage modern sands:* Università degli studi di Milano-Bicocca.

Loiseau, A., Asila, V., Boitel-Aullen, G., Lam, M., Salmain, M., & Boujday, S. (2019). Silver-based plasmonic nanoparticles for and their use in biosensing. *Biosensors, 9*(2), 78.

Lowry, S. R., Huppler, D., & Anderson, C. R. (1985). Data base development and search algorithms for automated infrared spectral identification. *Journal of Chemical Information and Computer Sciences, 25*(3), 235–241.

Mansur, A. A., Mansur, H. S., Caires, A. J., Mansur, R. L., & Oliveira, L. C. (2017). Composition-tunable optical properties of Zn x Cd (1– x) S quantum dot–carboxymethylcellulose conjugates: towards one-pot green synthesis of multifunctional nanoplatforms for biomedical and environmental applications. *Nanoscale Research Letters, 12*(1), 1–18.

Manuel, A. P., & Shankar, K. (2021). Hot electrons in TiO2–noble metal nano-heterojunctions: fundamental science and applications in photocatalysis. *Nanomaterials, 11*(5), 1249.

McPhail, D. (2006). Applications of secondary ion mass spectrometry (SIMS) in materials science. *Journal of Materials Science, 41*(3), 873–903.

Metzger, W., Ahrenkiel, R., Dippo, P., Geisz, J., Wanlass, M., & Kurtz, S. (2005). Time-resolved photoluminescence and photovoltaics:

National Renewable Energy Lab. (NREL), Golden, CO (United States).

Mishra, R. K., Zachariah, A. K., & Thomas, S. (2017). Energy-dispersive X-ray spectroscopy techniques for nanomaterial *Microscopy Methods in Nanomaterials Characterization* (pp. 383–405): Elsevier.

Myroshnychenko, V., Nelayah, J., Adamo, G., Geuquet, N., Rodríguez-Fernández, J., Pastoriza-Santos, I., Zheludev, N. I. (2012). Plasmon spectroscopy and imaging of individual gold nanodeca-hedra: a combined optical microscopy, cathodoluminescence, and electron energy-loss spectroscopy study. *Nano Letters, 12*(8), 4172–4180.

Nasse, M., Ratti, S., Giordano, M., & Hirschmugl, C. (2009). Demountable liquid/flow cell for in vivo infrared microspectroscopy of biological specimens. *Applied Spectroscopy, 63*(10), 1181–1186.

Nastasi, M., Mayer, J. W., & Wang, Y. (2014). *Ion beam analysis: fundamentals and applications*: CRC Press.

O'Connor, D. (2012). *Time-correlated single photon counting*: Academic press.

Pintori, G., & Cattaruzza, E. (2021). XPS/ESCA on glass surfaces: a useful tool for ancient and modern materials. *Optical Materials: X*, 100108.

Poehlman, F. S. W. (1980). *The development of a medium energy ion reflection spectrometer and some problems associated with its application to materials analysis*.

Riaz, T., Zeeshan, R., Zarif, F., Ilyas, K., Muhammad, N., Safi, S. Z., Rehman, I. U. (2018). FTIR analysis of natural and synthetic collagen. *Applied Spectroscopy Reviews, 53*(9), 703–746.

Shipp, D. W., Sinjab, F., & Notingher, I. (2017). Raman spectroscopy: techniques and applications in the life sciences. *Advances in Optics and Photonics, 9*(2), 315–428.

Shirley, B., & Jarochowska, E. (2022). Chemical characterisation is rough: the impact of topography and measurement parameters on energy-dispersive X-ray spectroscopy in biominerals. *Facies, 68*(2), 1–15.

Stevie, F. A., & Donley, C. L. (2020). Introduction to x-ray photoelectron spectroscopy. *Journal of Vacuum Science & Technology A: Vacuum, Surfaces, and Films, 38*(6), 063204.

Susi, H., & Byler, D. M. (1986). Resolution-enhanced fourier transform infrared spectroscopy of enzymes *Methods in enzymology* (Vol. 130, pp. 290–311): Elsevier.

Sutton, A. T., Kriewall, C. S., Leu, M. C., & Newkirk, J. W. (2016). *Powders for additive manufacturing process: characterization techniques and effects on part properties*. Paper presented at the 2016 International Solid Freeform Fabrication Symposium.

Szymanski, H. A. (2012). *Raman spectroscopy: theory and practice*: Springer Science & Business Media.

Taglauer, E. (1991). Ion scattering spectroscopy *Ion spectroscopies for surface analysis* (pp. 363–416): Springer.

Tang, Y., Lu, Y., & Luo, G. (2016). Controllable hydrothermal conversion from Ni-Co-Mn carbonate nanoparticles to microspheres. *Crystals, 6*(11), 156.

Tkachenko, N. V. (2006). *Optical spectroscopy: methods and instrumentations*: Elsevier.

Vandenabeele, P. (2013). *Practical Raman spectroscopy: an introduction*: John Wiley & Sons.

Weigel, S., Gehrke, M., Recknagel, C., & Stephan, D. A. (2021). Identification and quantification of additives in bituminous binders based on FTIR spectroscopy and multivariate analysis methods. *Materials and Structures, 54*(4), 1–9.

Weyland, M., Yates, T. J., Dunin-Borkowski, R. E., Laffont, L., & Midgley, P. A. (2006). Nanoscale analysis of three-dimensional structures by electron tomography. *Scripta Materialia, 55*(1), 29–33.

Wirth, R. (2009). Focused Ion Beam (FIB) combined with SEM and TEM: advanced analytical tools for studies of chemical composition, microstructure and crystal structure in geomaterials on a nanometre scale. *Chemical Geology, 261*(3–4), 217–229.

Wright, J. C. (1980). Double resonance excitation of fluorescence in the condensed phase—an alternative to Infrared, Raman, and fluorescence spectroscopy. *Applied Spectroscopy, 34*(2), 151–157.

Young, V. Y., Hoflund, G. B., & Miller, A. (1990). A model for analysis and quantification of ion scattering spectroscopy data. *Surface Science, 235*(1), 60–66.

# 5

# STRUCTURAL
# CHARACTERIZATION
# TECHNIQUES IN DETAIL

## 5.1 X-RAY DIFFRACTION TECHNIQUE (XRD)

XRD is one of the most often employed techniques for describing NPs. The XRD reveals the crystal structure, phase composition, lattice parameters, and crystal grain size. The latter parameter is computed via the Scherrer equation in addition to the widening of the strongest peak of an XRD measurement for a specific sample. The XRD technique has an advantage over other methods with respect to production of statistically representative, volume-averaged results. After drying the necessary colloidal solutions, these procedures are routinely used on samples that are in powder form. The composition of the particles can be determined by comparing the position and intensity of the peaks with the reference patterns in the International Centre for Diffraction Data (ICDD, formerly known as the Joint Committee on Powder Diffraction Standards, JCPDS) database. For particles less than 3 nm, the XRD peaks are too broad thus making it inappropriate for amorphous materials.

Other than the experimental broadening, particle/crystallite size and lattice stresses were the main causes of the broadening of XRD peaks. Even though the particle has a single domain, the XRD-derived size is often bigger than the so-called magnetic size because there are smaller domains in a particle where all moments are aligned in the same direction. (X-ray tomography is also frequently utilized in other industries such as metallurgy and materials research.) XRD is a non-destructive technique although it depends on the fact that X-rays are a form of light with wavelengths in the order of nanometers. When X-rays scatter from a material, a pattern of higher and lower intensity can result while this can cause interference. X-ray radiography and tomography are fundamentally dissimilar processes. The principle behind tomography is that some substances absorb X-rays more intensely than others. For instance, X-rays are more strongly absorbed by bone or malignancies than by muscle or fat. This makes the transmitted image an extremely useful tool for clinicians since it gives a direct image of the structure inside the body or object (usually at length scales of a millimeter or above).

As opposed to this, XRD creates a diffraction pattern that gives information about the internal structure of the sample while not resembling the underlying structure. An X-ray beam is directed at the sample, and the scattered intensity is calculated as a function of the outgoing direction. Conventionally, the angle between the directions of the incoming and leaving beams is referred to as $2\theta$. When Bragg's Law is met, constructive interference (higher scattered intensity) is seen for the most basic sample which is made up of sheets of charge separated by a distance d:

$$n\lambda = 2d \sin \phi$$

Here n is an integer (1, 2, 3...), $\lambda$ is the wavelength of the X-ray beam, and $\theta$ is half the scattering angle $2\theta$ as shown above.

Single crystallography involves growing a superior single crystal and arranging it in various X-ray beam orientations. The diffraction patterns that are produced can resemble. While the intensities can be used to determine the atomic locations within each unit cell, the position of the spots provides information on the symmetry and size of the crystal lattice. The diameters of crystallites as well as microscopic stresses and flaws can occasionally be determined by analyzing the shapes and widths of individual peaks.

Although they are the most challenging, single-crystal measurements typically provide more information than other XRD techniques. To gather the data required for a complete crystallographic determination, numerous measurements must be taken at various sample orientations because growing high-quality single crystals is at best challenging and frequently impossible.

The sample which frequently takes the shape of a finely crushed powder is made up of several crystallites rather than just one single crystal. The pattern is now made up of concentric rings with the same scattering angle $2\theta$ as a single spot would have had in a single crystal pattern as opposed to the pattern of sharp spots.

Two complementary applications of powder diffraction are most frequently used:

1. A substitute for single-crystal diffraction. A powder sample can be prepared significantly more quickly than a single crystal. As long as crystal structures are relatively small and there isn't too much peak overlap, they can still be solved using this method even though important information is lost during the "powder averaging" process that converts sharp points into rings. The Rietveld refinement technique is frequently used to identify the most likely crystal structure that is responsible for the observed pattern. Similar to single-crystal diffraction, it is occasionally possible to identify microscopic stresses and flaws as well as specifics of crystallite sizes by examining the shapes and lengths of individual peaks.

2. In mineralogy, phase identification is most frequently utilized. A mineral or clay sample will frequently have a blend of many crystal phases. The phase or phases present in nanomaterials can then be ascertained by comparing the "fingerprint" of a powder diffraction pattern to a database of known patterns.

The single crystal and powder techniques are both superior to the fiber diffraction method. The sample is often an extruded fiber with cylindrical averaging about a well-defined crystal axis aligned along the fiber axis (also known as the "meridian").

The 1953 discovery of the DNA structure is a well-known illustration of this method. It was difficult to grow true single crystals, and it was also difficult to analyze the data from single crystals at the time but the additional orientation of the diffraction pattern brought about by the fiber geometry was sufficient to determine the DNA molecule's helical structure. When researching long-chain compounds like DNA or columnar structures like discotic liquid crystals, fiber diffraction is frequently used.

The closely related techniques of grazing incidence diffraction (GID) also known as grazing-incidence X-ray scattering (GIXS) and X-ray reflectivity (XR) take advantage of the fact that the reflectivity is greatly increased when the beam of X-rays impinges on a surface at a very low incident angle, and when the beam penetrates the surface only a short distance. Therefore, this method is perfect for determining the characteristics of thin films or multilayers on substrates made of solids or liquids. A typical GID measurement holds $\alpha i$ constant while measuring intensity as a function of $2\theta$. It is possible to determine the 2-D crystal structure within the film's plane by analyzing the resulting intensity profile. A typical X-ray measurement involves fixing $2\theta$ at zero and measuring the reflected intensity of X-rays as a function of $\alpha i$. The thickness of the layer (or layers in a multilayer film) can be determined from the resultant intensity profile and in some situations; the electron density profile within each layer can be inferred.

### 5.1.1 Small-angle X-ray scattering (SAXS)

The term "small-angle X-ray scattering" (SAXS), often known as simple small angle scattering (SAS), designates studies with a tiny scattering angle $2\theta$, typically less than $10°$ (Li, Senesi, & Lee, 2016). According to Bragg's Law, this suggests that the objects being probed have a rather broad length scale, often between 3 and 100 nm. This method was historically used to investigate relatively big "objects" that were scattered in a medium such as proteins that were dissolved in an aqueous solution, colloidal particles, micelles, or voids in porous media.

With repeat distances far greater than a single molecule, self-assembled systems such as block copolymers that have periodic order have been studied using SAXS more recently. A cubic structure that is 20 nm across or more is created when many tens of molecules self-assemble into spheres. In this instance, the atomic positions are highly disordered but the positions of the spheres are long-range ordered. Instrumentation designed for scattering at tiny angles is needed for such systems while crystallographic analysis methods are more appropriate for the analysis process.

When we consider the production of X-rays in the XRD machine, there are a variety of methods for producing a beam of X-rays.

1. X-ray tube – The simplest and earliest method is still occasionally employed. X-rays are produced when a metallic target is struck by an electron beam. The heat that the electron beam causes to be emitted into the target limits the X-ray beam's intensity.
2. Anode-rotating X-ray generator – This type of X-ray tube which became widely accessible in the 1970s solves the heat

loading issue by swapping the fixed target for a revolving cylinder that is water-cooled internally. However, there are both literal and figurative costs: the engineering specifications are much more severe and rotating anode generators are prone to failure and require frequent maintenance. As a result, a significant increase in X-ray intensity is made possible.

3. Synchrotron – In comparison to the tube sources mentioned above, a synchrotron X-ray source emits radiation from a relativistic beam of electrons (or positrons) that has been accelerated by a magnetic field. When compared to the beam produced by the sources mentioned above, the resultant beam is typically many orders of magnitude more powerful. However, such a beam can only be produced by a sizable centralized facility and most users are forced to travel great distances and schedule their usage well in advance. Due to this, investigations requiring extremely high intensities or other unique circumstances should use synchrotron sources instead of tube/rotating anode/micro-focus sources which can be operated at the user's home institution. In addition to the Advanced Photon Source and the National Synchrotron Light Source in the US, other significant synchrotron sources include the Photon Factory in Japan, the European Synchrotron Radiation Facility in France, the Diamond Light Source in Britain and the European Synchrotron Radiation Facility.

The process of creating a monochromatic beam; one with a limited range of wavelengths using a crystal monochromator involves introducing a high-quality single crystal made of silicon or germanium into the beam and separating out only the parts of the beam that adhere to Bragg's Law (Verbeni et al., 1996). However, this Bragg reflection from a crystal can be employed as a collimation method for a beam that is already mostly monochromatic. The degree of collimation and spectrum selection are influenced by both the incoming beam's properties and the crystal's level of perfection.

In SAXS method, the electronic density fluctuations within a material are measured by using the elastic scattering of X-rays at very low angles ($0.1e^{-10}$). SAXS measurements use X-rays at high scattering angles so they can examine length scales from angstroms to micrometers and allow for the direct determination of structural information for systems with large-scale random density fluctuations (changes in electron density) in real time. This allows for the analysis of the internal structure of disordered systems.

1-D average intensities are commonly generated via circularly averaging 2-D scattering patterns from SAXS observations. Information on the size, shape, and spatial configurations of the nanomaterials is available from the scattering patterns. The parameters of interest and any prior information about the samples will determine the data analysis techniques to be utilized. The zero angle scattered intensity and the root mean square (RMS) radius which are related to the size and shape of the scattering particle can be computed with high precision by carefully examining a high-quality scattering profile using Guinier's approach. Broad species distributions make it challenging to distinguish between individual populations; however, many populations of sufficiently different sizes can produce distinct patterns in the scattering curve. Only when the scattering intensity is measured on an absolute scale, it is possible to determine physical attributes like molecular weight, particle volume, and specific surface area (see next section). A precise sample concentration must be known in order to determine the molecular weight from the scattering data because the analyte's concentration and the molecular weight are proportional to the scattering signal. Though form factors can also be obtained from SAXS experiments, assumptions about shape (spheres) are frequently utilized for in situ monitoring of nanomaterial production.

Similar to the procedures used for analysis, the requirements for sample preparation to clarify parameters of interest will alter. Sample preparation is crucial for successful SAXS measurements of nanostructures since the scattering signal needs to be obtained from monodispersed particles of the same composition in order to get exact structural characteristics. Sample aggregation should be prevented and the sample to be measured must be extremely pure. However, as was already indicated, a benefit of SAXS over Cryo-electron microscopy (EM) involves its capacity to more accurately statistically quantify the product distributions with little sample preparation (Minelli et al., 2018). The requirement of known shape factors which would need to be determined with coupled EM measurements is an inherent disadvantage of SAXS measurements for determining the nanomaterial distributions. However, the use of orthogonal techniques for validation is a general requirement for validating many nanomaterial measurements as outlined in this section.

## 5.2 ELECTRON MICROSCOPY

The core of electron microscopy and the family of microanalysis characterization techniques along with the signals created by the scattering of energetic electrons by atoms within a specimen offer a lot of information about its structure and composition. Compared to optical approaches, these methods can produce substantially better spatial resolutions.

Microscopy is possible because of the incident radiation's significantly shorter wavelength. For instance, visible light has a wavelength of 390–700 nm, whereas 30 keV electrons have a wavelength of 41 pm (Minelli et al., 2018).

The characterization of nanomaterials can benefit greatly from the spatial resolution and chemical sensitivity of the electron beam techniques. However, the incident electron beam, however, has the potential to drastically change or entirely destroy the material being studied because of the beam-induced ionization and sputtering. Moreover, these methods must be used in high vacuum (HV) environments unless when using specialist equipment (Bugnet, Overbury, Wu, & Epicier, 2017). This can affect samples that are sensitive to the gaseous environment in which they are housed physically and chemically thus limiting the types of samples that can be evaluated.

### 5.2.1 Scanning electron microscopy (SEM)

Depending on the application, an electron gun assembly in the SEM produces a stream of electrons with primary energy that typically varies between 500 and 40 keV. Electromagnetic lenses concentrate the beam to a small point which is then raster-scanned across the specimen's surface using scan coils to regulate the beam's position. By serially collecting one or more of the signals created by the interaction between the incident electrons and the atoms within the specimen, a spatially resolved image is created. The analyst must carefully analyze the issue at hand and select the appropriate experimental conditions in order to guarantee that the data acquired is as relevant as possible. The SEM may access a wide variety of signals linked to the structural and chemical makeup of the specimen. It is frequently used to evaluate micro- and nanoscale topography and morphology using the secondary electron (SE) signal which consists of low-energy (50 eV) electrons ejected from the sample by the inelastic scattering of the source beam. Only SEs produced near the specimen surface and within a short distance of the beam point can escape due to their low energy and contribute to the gathered picture signal. As a result, the lateral and depth resolutions of SE images can be extremely high (1 nm).

Chemical contrast can be obtained by gathering the backscattered electron (BSE) signal, or the BSE signal. In this situation, there is no appreciable energy loss as the primary beam electrons are elastically reflected back toward the direction of incidence. The intensity at any given pixel in a BSE image is inversely related to the average atomic number present at that location in the specimen because more heavy atoms strongly backscatter electrons than do lower mass atoms. Compared to secondary electrons (SE) imaging, topographic contrast and edge effects are lessened in this image mode enabling a more quantitative interpretation of the contrast. For nanomaterial characterization, this imaging mode is incredibly helpful for boosting the contrast between heavy metal NPs from a lower Z matrix such as those found in supported metal catalysts or particles embedded in organic matter. One of SEM imaging's main benefits is the extraordinarily wide depth of field it can achieve. This has a resolution of tens of microns allowing simultaneous imaging of all regions of a complicated, topographic object. For instance, supported metal catalysts contain nanomaterials spread across larger substrate particles. The latter are known to aggregate regularly to build larger, 3-D networks that are highly complex and have a high degree of topography. These networks can be immediately imaged via SEM due to the great depth of field thus making it easier to assess the 3-D structure of the agglomerates, degree of agglomeration, and other factors.

A second column is commonly added to the SEM to produce a concentrated beam of heavy ions that can be utilized to remove material locally and expose internal features. The focused ion beam (FIB) source can be any low melting temperature metal or gaseous plasma. In the former, gallium is the most often utilized source material whereas xenon is frequently employed in FIBs with a plasma source. Characterization of nanomaterial distributions embedded inside a matrix is the primary use of the FIB for nanomaterial investigation. This is performed by integrating a "slice-and-view" technique in which the ion beam removes a small portion of material and the newly revealed underlying face is then evaluated using any of the previously covered SEM signals (Guehrs et al., 2017). This removal/

analysis operation can be repeated numerous times to obtain the elemental composition and 3-D dispersion of nanomaterials.

## 5.2.2 Transmission electron microscopy (TEM)/ high resolution (HR-TEM) with selected area electron diffraction (SAED)

Transmission electron microscopy (TEM) is a microscopy technique that takes advantage of the interaction between a thin sample and a uniform current density electron beam with energies typically ranging from 60 to 150 keV (Reimer, 2013). A portion of the electrons in the electron beam is transmitted when it hits the sample while the remainder is dispersed either elastically or inelastically. Size, sample density, and elemental makeup are just a few of the variables that affect how much of an interaction there is. The information gathered from the sent electrons is used to construct the final image. As was made evident in the preceding sections, the size and morphology of NPs determine their specific set of physical characteristics including their interaction with biological systems and their optical, magnetic, electrical, and catalytic activities. Since TEM offers direct views of the sample as well as the most precise estimation of the homogeneity of the NPs, it is the method most frequently used to analyze NP's size and the form. However, there are several restrictions that must be taken into account while employing this method such as the challenge of measuring a large number of particles or inaccurate images caused by orientation effects. Other methods that analyze more NPs can produce more accurate findings when describing extremely homogeneous samples such as SAXS for bigger, spherical NPs or XRD by utilizing the boundary of the XRD reflections and the Scherrer formula. To ensure sample uniformity, a prior analysis must be carried out.

In addition to their size and morphology, NP qualities also depend on other elements, such as interparticle distance. For instance, when two metal NPs are brought together, the pairing of their plasmons occur causing their plasmon band to move to the red region and their color to change. To characterize the NPs aggregation using TEM for various biomedical applications including sensing and diagnostics where the aggregation depends on the presence of a biomarker or analyte. In the application of therapy, the aggregation of NPs increases the therapeutic effect of the NPs whereas in, imaging, the aggregation enhances the response signal. Extra caution should be used when preparing samples to ensure accurate results because a poor methodology could change the sample or produce artifacts such as aggregation during the drying of the colloid solution.

The methodical assemblage of several nanocrystals results in new multifunctional structures that combine the advantages of the constituent parts as well as give rise to the emergence of novel and intriguing capabilities. One method used to characterize the creation of various super-lattice nanocomposites which can be isostructural to a number of atomic crystal systems is TEM. The final structure and composition of these new 3-D arrays which are made of various NPs (such as metals, magnetic NPs, and quantum dots) can be modified by adjusting the colloidal surface charge or directional bonding with DNA.

The scientific community has begun to see NPs as dynamic systems in the last few years where their properties and structure can change over time as they interact with their environment. To improve their performance in a variety of applications, it is crucial to characterize their dynamic alterations. For instance, Ag NPs have been shown to assemble in the presence of sunshine which reduces their cytotoxicity. When exposed to sunlight, the NPs produce nanobridges among themselves according to TEM imaging. The morphological modifications and surface sulfidation had an impact on the dissolving rate of the NPs which reduced their toxicity. Additionally, the biodegradation of the NPs polymeric covering by bacteria has been investigated using TEM and DLS. Colloidal aggregation results from the loss of the particle coating which reduces their mobility.

Moreover, the growth of NPs cannot be studied using conventional TEM. However, it can be used to describe how colloids arise from solid predecessors. For instance, copper NPs' growth kinetics has been imaged by TEM. TEM with high resolution (HR-TEM) employs phase-contrast imaging which combines both transmitted and scattered electrons to create the image. To use the scattered electrons, HR-TEM requires a bigger objective aperture than the conventional TEM imaging. The method with the best resolution ever created and phase-contrast imaging enables the observation of arrays of atoms in crystalline formations. While conventional electron microscopies can provide a statistical analysis of NP morphology, they do not have enough resolution to photograph the single particle crystal structure. HR-TEM offers critical information on the NP structure. As a result, HR-TEM has emerged as the most frequently used method to describe the internal structure of NPs.

Figure 5.1 displays a high-resolution image captured on a traditional TEM utilizing reasonably high beam energy of 400 keV. Since a beam with greater energy produces a shorter electron wavelength, this technique has been the most effective way to increase the 'point' resolution for a long period of time. A single AgI NP resting on a carbon-based support may be seen in the image. The fast Fourier transform (FFT) of the image provided as an inset demonstrates that the NP is oriented such that the electron beam is parallel to the *110* zone axis. The NP has a well-defined truncated octahedral shape with *100* and *111* facets. When examining the structure of low atomic number material, phase contrast exhibited in this image is useful. It is obvious from this image that the carbonaceous support immediately beneath the NP in the picture is graphitic in character with three to four layers of graphite that appear to be coherent with the NP's *200* planes. However, phase-contrast images, especially those from non-aberration adjusted microscopes also need to be evaluated carefully because rapid contrast reversals can happen with changes in defocus and specimen thickness (Thomas et al., 2013).

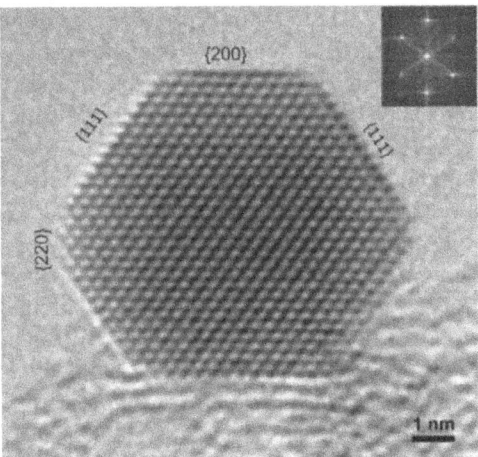

**Figure 5.1** High resolution TEM image of an AgI NPs recorded parallel to the <110> zone axis, exhibiting a truncated octahedral morphology. The planes seen in the figure are labelled. The inset shows the FFT of the NP (Thomas, Midgley, Ducati, & Leary, 2013).

Additionally, although having identical optical qualities, single crystal and polycrystalline anisotropic NPs can be distinguished using HR-TEM. Additionally, HR-TEM enables the characterization of structural transitions such as the temperature change in the iron-platinum NPs from disordered face-centered cubic to ordered L10. Despite the fact that HR-TEM is a powerful approach, it is important to note that it is not always possible to characterize NPs using this method. There may be directions where the atoms are not well aligned due to the random orientation of the crystals with respect to the electron source leading to complicated pictures that cannot be used to directly characterize the structure.

### 5.2.2.1 Liquid TEM

As was already established, a key element of TEM is the vacuum system which guards against possible filament damage and lessens scattering of the electron beam. Since liquids could impair the vacuum, traditional TEM imaging has only been employed on solid and dried samples. Thus, for many years, there has been a lack of attention paid to the nanoscale characterization of solid-liquid systems. L. Marton observed biological samples trapped between thin aluminum foils in the 1930s which marked the beginning of early attempts to characterize liquid samples. However, until recent years, when the nanofabrication of sealed liquid cells was established, no substantial breakthrough had been made due to the technological issue of maintenance of vacuum and prevention of the liquid from evaporation. In 2003, Frances M. Ross and colleagues created a TEM liquid cell employing epoxy sealed silicon nitride (SiN) membranes. These membranes contained the liquid sample and were electrons transparent thus maintaining the vacuum in the microscope.

By vitrifying materials at cryogenic temperatures, cryo-electron microscopy (cryo-TEM), a subclass of TEM, enables the observation of nearly unmodified samples in their natively frozen environment. Dubochet, Frank, and Henderson received the 2017 Nobel Prize in Chemistry recently for their creation of cryo-electron microscopy for the high-resolution structural identification of biomolecules in solution. The samples are normally frozen using liquid nitrogen. Due to the absence of variables (such as staining and sample storage in non-physiological conditions) that could change the conformation or assembly of the sample's molecules, this technique is frequently utilized in molecular biology and colloid chemistry.

Typically, commercial automated plunge freezers are used to vitrify the liquid samples. These freezers freeze water solutions by rapidly lowering their temperature and preventing the water molecules from reorganizing into long range structured crystal lattices. As a result, an amorphous state resembling the native liquid is produced. This amorphous condition is achieved by plunge freezers in four steps: (1) addition of the liquid sample on the copper grid with carbon coating; (2) draining the excess liquid to create a thin film; (3) submerging the grid in liquid $N_2$ to freeze; and (4) keeping the vitrified sample in a storage container with liquid $N_2$.

Diffraction patterns obtained by electron microscopes can be analyzed to provide information about the structure of the specimen. This brief review covers some of the technical aspects of using electron diffraction patterns for structural analysis. The technique of Selected Area Electron Diffraction (SAED) makes use of diffraction from a specific region of the specimen.

The objective lens collects electrons that emerge from the specimen and magnifies them to focus all of the electrons that emerge from a single point on the image, regardless of the angle at which they emerge. As electrons travel from the specimen to the image, there is a plane known as the back focal plane or diffraction plane where all electrons that emerge at a single angle are focused onto a single point regardless of where they emerge on the specimen. To put it another way, electrons with a specific momentum vector are focused at a specific point. Lenses further down the column can be set to magnify either the image of the specimen or the diffraction pattern depending on whether the object planes for the lenses are the image plane or the back focal plane of the objective. As a result, each lens has planes where the image or diffraction pattern is focused; these planes are conjugate to either the image or diffraction plane, respectively. An aperture will transmit electrons from the portion of the image that is inside its hole and block electrons from the portion of the image that is outside its hole on the solid section of the aperture if it is inserted into one of the planes conjugate to the image plane. Only the data from the portion of the picture contained within the aperture hole is transferred to the detector regardless of whether the microscope is in the imaging mode or diffraction mode. The appearance of a selected area image is just that of an ordinary image with a black circle over it. Since the scattering from parts of the specimen that are not immediately of interest is removed, a diffraction pattern of selected region has less background than if the aperture had not been placed.

Diffraction data from a specific area of the specimen can be obtained by appropriately determining the size and placement of the designated area aperture (Lendaris & Stanley, 1970). This can be used to choose a crystalline region while excluding an amorphous region or obviously the other way around. It can also be used to choose a region with a specific orientation or phase while excluding regions with other orientations or phases.

### 5.2.3 Environmental transmission electron microscopy (ETEM)

The discussion up to this point has only covered conventional methods of electron microscopy, all of which are conducted under HV or ultra-high vacuum (UHV) conditions to reduce electron scattering by surrounding gas molecules. However, exposure of the specimen to such a setting may have unexpected impacts on its chemistry and structure, particularly if it is in the form of a highly reactive nanoparticulate. Due to this, methods for conducting electron microscopy and microanalysis in gaseous or liquid environments have been developed. By separating the area of the instrument containing the specimen from the rest of the instrument, this is achieved. In this approach, the sensitive parts of the instrument such as the electron cannon are safeguarded and the majority of the beam flight path occurs in a HV (i.e., low scattering) environment. The environmental SEM and TEM (often referred to as ESEM and ETEM, respectively) use a multistage differential pumping aperture as one method. The discovery of these methods has improved our knowledge of the mechanisms that drive carbon nanotube (CNT) growth and the alterations that oxide nanomaterials go through during catalytic reactions. The specimen can also be isolated by placing it a tiny vessel within the TEM column or SEM chamber. The vessel also known as a "wet cell" is made in such a way that the incident probe and desired signal can travel through a thin but water-impermeable membrane (Nishiyama et al., 2010). The ESEM/ETEM is often employed for regulating the gaseous environment around the particles whereas this so-called "wet-cell" method is most frequently utilized for imaging particles in liquid. When the "wet-cell" is sealed and introduced into the instrument, it can be completely isolated which prevents any liquid exchange. As an alternative, the outside of the instrument and the interior of the cell can be connected through fluid flow channels.

## 5.2.4 Scanning probe microscopy (SPM)

SPM is a technique for sample surface observation that relies on a physical probe rather than light to examine a specimen. This offers a lot of data that is not possible with light microscopy. SPM has an atomic resolution that consistently outperforms even cutting-edge methods like SEM and TEM in resolving sub-nanometer features.

This tiny physical probe with an extremely pointed nanoscale tip that engages with the sample is the brain of any SPM. SPMs come in a variety of forms and what sets them apart is how they interact with the material at the tip.

Scanning Tunneling Microscopy (STM), which was first created in 1982, is said to be the original type of SPM. The physical sensing probe used in STM is a tiny wire with an extremely sharp tip that has been cut or engraved. This tip is scanned across the sample surface by the SPM raster's piezoelectric scanner. By monitoring the tunneling current between the tip and the sample, a STM can detect the surface. It is an extremely responsive sensor because this current ranges from a few nanoamperes to several picoamperes and fluctuates exponentially with the tip to sample distance. However, this interaction also establishes one of the main drawbacks of STM: the requirement that the sample be conducting. The atomic force microscope (AFM) was created in 1986 as a solution to this constraint. The wire probe of the STM is replaced by a micro-machined probe in the AFM which is commonly created by photolithography and etching silicon wafers.

### 5.2.4.1 Atomic force microscopy (AFM)

AFM is a potent scanning probe microscopy technology that may be used to investigate any solid surface at extremely high spatial resolutions and provide details on morphology and surface characteristics. A laser detector locates the tip of a cantilever that is positioned on a stage. The tip is brought close enough to a surface such that the probe can "detect" the attractive and repulsive forces that are present in the sample. 2-D piezoelectric scanners sweep the probe laterally. The potential energy difference between the cantilever tip and the sample surface causes the probe to move vertically as it scans across the specimen. The tip to sample force is employed by the AFM as the imaging signal depending on the mode of operation and a feedback loop is used to keep the force on the tip constant by modifying the sample height.

A 3-D map of the sample surface is created once the changes are recorded and translated. Resolution of AFM images is consistently at angstrom levels in the z-dimension (height) but is constrained in the lateral dimensions by the curvature of the tip (typically 10 nm or more). To obtain information regarding the strength and geometry of chemical bonds forming between individual atoms, a direct relationship between the experimental signal and the underlying physical forces has been developed. Additionally, the analysis can be done in natural light.

The height, size, shape, aspect ratio, and surface morphology of individual particles can all be described using AFM. The greatest height of a particle is used to quantify particle size in AFM studies and 3-D analytic tools can be used to determine particle size distributions. Nanomaterial diameter, volume, and surface area can also be computed using the three dimensions of data presented in the AFM pictures. The probe geometry has an impact on these measurements and for particle size measurements; the cantilever is frequently calibrated in the z-dimension. The cantilever geometry has a significant impact on a number of other parameters including the particle volume and circumferences. Characterization is typically challenging for particles placed on uneven or curved substrates. For AFM imaging, a substrate must have particles that are tightly bonded to it, evenly disseminated throughout it and with a roughness that is less than the size of the nanomaterials. However, many different combinations of nanomaterials, substrates and adhesives have been shown to be appropriate for AFM and each one needs the right surface treatments. Literature has covered a number of sample and surface treatments for characterization of nanomaterials (Fendler, 2008).

A millimeter-scale "chip" that is attached to a cantilever that is typically rectangular and less than 200 micrometers long and a few tens of micrometers wide makes up an AFM probe. This cantilever has an extremely sharp point (the "tip") that extends down a few micrometers and ends at a little point with a radius of around 10 nm. When the tip of this cantilever hits the surface of the sample, it acts as a spring and bends. By reflecting a laser off the rear of the cantilever and monitoring the motion of the reflected spot using a split photodiode detector, the AFM can detect this bending. This cantilever-based AFM detector is incredibly sensitive to the tip to sample distance despite not using the same concept as STM.

SPMs' picture samples are created by slowly scanning in an orthogonal manner down across the image region while raster scanning the tip across the sample back and forth, line by line.

The SPM's usual picture area might range from a few nanometers to up to 100 micrometers. Therefore, conventional positioning technologies are inappropriate because they generate motion on a scale that is often orders of magnitude bigger than what is needed (Zhang, Zhang, & Luby, 2007). Instead, piezoelectric crystals are used for positioning samples for SPM. When biased with an electric field, these ceramic materials expand thus producing microscale motions.

### 5.2.4.2 Scanning tunneling microscopy (STM)

Without using light or electron beams, STM is an imaging technique used to acquire ultra-high resolution images at the atomic scale. Two IBM scientists named Gerd Binnig and Heinrich Rohrer created STM in 1981. They won the Physics Nobel Prize five years after their invention.

The first method in the larger category of SPM imaging modalities was STM. Researchers were able to record much more detail, down to the level of atoms and interatomic distance, at the time than with any other type of microscopy. Researchers were able to precisely map the 3-D topography and electrical density of states of conductive materials using these ultra-high resolution capabilities, and to even modify individual atoms on the surface of these materials. STM has significantly advanced the science of nanotechnology during the ensuing decades and continues to be crucial for both basic and applied research in a range of fields.

STM is a remarkable and an uncommon example of electron tunneling usage which is a quantum mechanical process in a real world practical application. The process of electrons passing through a barrier that initially seems impenetrable like throwing a ball against a wall, in this case, a tiny gap between the tip and surface is referred to as "tunneling." The ball will never tunnel through the wall according to the "classical paradigm" of physics, which characterizes this ball-wall interaction. Unlike a ball, electrons are more like a "fuzzy" cloud and can truly exist on both sides of the barrier at once. As a result, even though the barrier's energy is more than the total energy of the electron, there is a non-zero probability that the electron will cross it.

To conduct STM, a sharp conductive probe must be moved very close to the surface of a conductive object. The fuzzy electron cloud of the first atom of the tip and surface starts to overlap when the tip is sufficiently close enough to the surface (often less than 1 nm distant). The overlapping electron cloud drives electrons to the tunnel through the potential barrier from the tip to the surface thus creating a current when a bias voltage is applied between the tip and the surface in this configuration. This tunneling current varies exponentially with the tip to sample distance and is extremely sensitive to the distance between the probe tip and the surface. The intensity of the tunneling current maps the sample's electrical density of states as the tip scans the sample's surface line by line.

Constant height mode and constant current mode are the two distinct ways that the STM functions. When the sample surface is exceptionally smooth, the constant height mode is typically utilized. In this mode, the sample is swiftly raster scanned while the probe tip remains at a fixed height. Researchers can create an image of the electronic density of states of the sample surface, defects, frontier molecular orbitals, and more by observing changes in the strength of the tunneling current as a function of (x, y) position and bias voltage.

The constant current mode is the more widely used mode. By adjusting the gap between the tip and the surface in this mode, a feedback loop system maintains a constant tunneling current. In other words, if the tunneling current is greater than the target value, the feedback control system will distance the tip from the sample; conversely, if the tunneling current is lower than the target current value, the tip will be brought closer to the sample's surface. Researchers can assess a variety of properties such as surface roughness, flaws, the size and the conformation of molecules on the surface, using the resulting 3-D distance profile as a function of (x, y) position.

The STM was initially used to characterize the topology of various metals and to delineate the atomic structure of their surfaces. Researchers were able to discern atomic-scale properties of materials such as surface roughness, defects, and surface reaction mechanisms for the first time. Researchers could start to understand properties relevant to the fabrication of electronic components by investigating the atomic lattices of materials such as conductivity, distributions of frontier molecular orbitals, and their energies and reaction dependencies on crystal facet orientations, to name a few (Kalinin & Balke, 2010).

STM has been used for a variety of applications other than atomic-scale imaging over the years. It has been employed in the assembly and manipulation of individual atoms on a surface. This opened up new possibilities for nanotechnology such as the creation of nano-structures like quantum corrals and molecular switches. STM can also be used to create contacts on nanodevices by depositing metals (such as gold, silver, or tungsten) in a specific pattern. STM has also been used by researchers to induce chemical reactions and study the subsequent reaction mechanisms at the molecular level.

## 5.3 DYNAMIC LIGHT SCATTERING

A potent technique for characterization is the interaction of nanoma-terials and light. The energy source generates a wave vector which can change in response to the physical properties of the particle, such as size and shape, while the frequency remains constant or absorbs the energy. Because the strength of scattered light is related to particle size and inversely proportional to light wavelength, larger particles scatter more light than smaller ones, and shorter wave-lengths scatter more strongly than longer ones. A number of nano-material characterization techniques can then be used to measure the resulting signal.

DLS, also known as photon correlation spectroscopy or quasi-elastic light scattering, identifies time-dependent variations in par-ticle light scattering intensity using a digital auto-correlator. The fluctuations are caused by randomized Brownian motion and are inversely related to the rate of translational diffusion of the particle through the solvent. The Stokes-Einstein equation is then used to connect this diffusion coefficient to the particle's hydrodynamic radius, which roughly corresponds to the apparent size taken up by a solvated, tumbling molecule. It is one of the few non-destructive procedures for determining average hydrodynamic diameter and requires little sample preparation.

The scattering angle, particle size and shape, instrument optics, and measurement time all have an impact on the DLS scattering intensity. Furthermore, for irregularly shaped particles bigger than roughly half the laser wavelength, scattering intensity depends on the particle form factor. Due to the very nonlinear relationship between particle size and scattering intensity, it is still challenging to establish the true particle size distribution.

Extraneous light scattering occurs when very large particles, such as dust and nanomaterial aggregates, are present. These larger particles distort the apparent size distribution, which in some circumstances completely conceals a numerically dominant lower size fraction. They also decrease measurement reproducibility. Samples in highly poly-dispersed systems must therefore be fractionated or filtered before being subjected to DLS analysis (Filipe, Hawe, & Jiskoot, 2010). The concentration of the analyte should also be taken into account because metallic NPs have a low scattering intensity (50 nm) which makes detection and reproducibility challenging at low concentrations.

By measuring the scattering angle and light intensity of incident light, static light scattering (SLS), also known as laser diffraction employs the Mie theory to calculate the size of nanomaterials (Joosten, McCarthy, & Pusey, 1991). Although the Rayleigh-Gans-Debye (RGD) approximation is frequently used for diluted solutions where the solvent and nanomaterial have similar refractive indices. It has been demonstrated that it also corresponds well with orthogonal measurements of size that fall strictly outside of this circumstance. In general, orthogonal measurements are always recommended to verify the appropriate use of the RGD approximation or other formalisms to derive size. A solution containing a suspension of the target analyte for SLS is examined using a well-collimated, highly coherent beam of polarized light. The plane where the intensity and angular dependence of the subsequently scattered light are to be measured has a perpendicular electric field to the incident light. SLS can determine the structure and mass of molecules over a wide range of molecular weights and is one of the few absolute techniques

By measuring the angular dependence of the scattered light, the size of the analyte can be identified. The root mean square (RMS) radius which is a measure of particle size weighted by the mass distribution around its center of mass is used to determine size for SLS. The RMS radius depends on the internal mass distribution of the particle and can be connected to its geometrical dimensions if the particle's conformation (such as random coil, sphere, or rod) is known. Due to the fact that the intensity of light scattered by a particle is inversely proportional to its molar mass and concentration in solution, SLS is an effective method for detecting the presence and development of aggregates.

Similar considerations apply to sample preparation as they do for DLS where large particles may impede the identification of smaller particles and extremely low concentrations make it challenging to reproduce scattering intensities. Furthermore, for diluted solutions with negligible interactions between sample particles, solvent molecules, or one another, the RGD approximation is valid. Furthermore, it is assumed in the calculations that the solvent's refractive index and that of the particle are almost identical. Data analysis, instrument alterations, and/or the use of this approach as a detector after a fractionation phase can be used to make up for samples that cannot be processed in a manner that satisfies these requirements.

Mean Hydrodynamic Diameter (MHD) and the polydispersity index (PDI, which represents the square of the ratio between the absolute width and the mean of the DLS spectra distributions) of the liposomes were studied. The direct comparison between the two 45° and 90° chips was performed at FRR = 5 and at the lowest lipid concentration. As observed in Figure 5.2a, liposome Z-average MHD was of 92 nm with a wide PDI (0.35) for the 45° chips, while in the case of the 90° chips, the average size was apparently larger (120 nm), but with a lower PDI (0.27). Figure 4.1b represents a comparison between the Z-average MHD of the liposomes produced with the two chips at different Flow-Rate-Ratios (Zizzari et al., 2017).

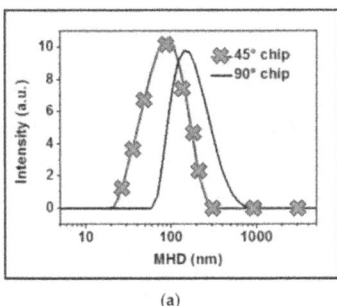
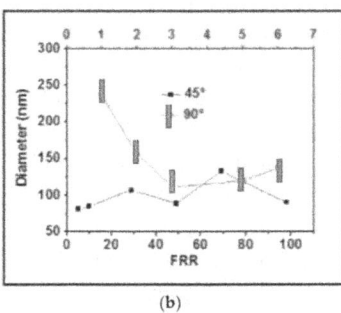

(a)  (b)

**Figure 5.2** (a) Dynamic light scattering (DLS) spectra of liposome solutions produced inside 45° and 90° chips at FRR = 5 and at a lipid concentration of 0.9 mg/mL; (b) plot of the Z-average mean hydrodynamic diameter (MHD) versus the FRR of liposomes produced at a lipid concentration of 0.9 mg/mL inside 45° and 90° chips. The black x axis is related to FRRs used in the 45° chips and the red one (upper side of graph) to FRRs used in the 90° chips (Zizzari et al., 2017).

## 5.4 X-RAY ABSORPTION SPECTROSCOPY (XAS)

The terms X-ray absorption fine structure (XAFS) and X-ray absorption near-edge structure (XANES) are collectively referred to as X-ray absorption spectroscopy. In both methods, atoms in a sample absorb X-rays while the radiation's energy is changed and each type of element results in a distinct pattern of absorption. The sample's X-ray absorption spectrum is represented by this plot of absorption against energy. Every element absorbs X-rays at a different wavelength; hence the important characteristic of these XAS approaches is that each measurement can be "tuned" to only seek for an element of interest by changing the X-ray source's wavelength. A synchrotron facility is necessary in this situation.

The powerful and continuously adjustable X-ray beams are required for XAS and it can be produced by synchrotron X-ray radiation. This enables the beam's energy to be transformed into the types of element that are particular to the target. An important approach for locating the components in unidentified samples is the element specificity. This makes the method extremely helpful for discovering hidden layers in materials that are otherwise unreachable for investigation especially when combined with the high penetration of X-ray radiation. These XAS techniques are excellent for examining environmental pollution because they can distinguish between chemical oxidation states and compounds containing harmful elements like arsenic, selenium, or chromium at the parts per million levels. As in XAFS technique, the arriving X-rays cause the target atoms to release electrons that interact with neighboring atoms to provide information about the target atom's environment. Contrarily, the XANES method gives information about the atoms being struck by the X-rays and is highly sensitive to their oxidation state and shape making it valuable for identifying the precise chemical species present.

One of the main benefits of XAS techniques is that they may detect solids, liquids, gases, mixes, and amorphous materials without the need for crystalline samples unlike diffraction techniques. The method is also non-destructive which makes it appropriate for delicate samples (like artwork or historical relics) and samples with limited resources. Even trace elements in a sample can be found using the PETRAIII X-ray source. In general, the extremely quick analysis allows for in situ probing of structural changes occurring during chemical events (such as catalytic reactions).

Materials science, chemical research and biochemical research are some of the major fields where XAS is used frequently for the identification of metals in a sample (for example, metals, NPs, or in the active sites of metalloproteins/metalloenzymes) and their oxidation states. Aside from analyzing static samples such as metals during alloying and catalysts during processes, XAS is particularly effective for researching dynamic samples due to its speed. The study of NPs and catalysts have found particular use for XAFS. Other intriguing applications of the XAFS approach include the characterization of impurities such as those identified by their X-ray fluorescence or the measurement of chemical valence states in high temperature superconductors.

There are two regions in XAS:

- (XANES/NE-XAFS: This region features prominent resonance peaks and includes X-ray energies that are closest to the absorption edge (about 100 eV around the edge). In general, the area is sensitive to local atomic states like symmetry and oxidation states.
- Extended fine structure (EXAFS): This region has features that occur after the XANES region and go up to or past the absorption edge, about 1,000 eV or more. EXAFS is characterized by mild oscillations in the recorded signal and is brought on by the scattering of the expelled electron by the atoms in the immediate vicinity. Bond lengths and chemical coordination environments of nearby atoms can be measured via EXAFS experiments.

X-ray absorption (XAS) measurements were collected at the Ir L3 absorption edge (11.215 keV) and are shown for three selected pressures in Figure 5.3. These spectra can be decomposed into three main contributions: (i) a sharp, atomic-like "white-line" that can be assigned to the $2p_{3/2} \rightarrow 5d$ electronic transition, (ii) a step-like edge associated with $2p_{3/2} \rightarrow$ continuum electronic excitations, and (iii) smaller oscillations (fine structure) that result from the photoelectron backscattering from neighboring atoms (Monteseguro et al., 2019).

## 5.5 THE BRUNAUER-EMMETT-TELLER (BET)

The surface of a nanomaterial contains a significant portion of its atoms, and it is these atoms that frequently control the particle's

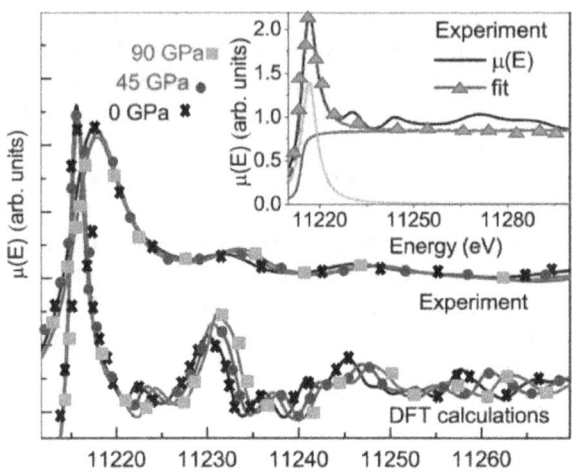

**Figure 5.3** Experimental (simulated) x-ray absorption spectroscopy (XAS) spectra at ambient pressure, at 45 GPa and 90 GPa in black line, red line and blue line, respectively. Te Lorenztian+arctang ft of the White Line (experimental spectrum at room pressure) is represented in the inset. Red dash line is the total ft of the WL, the blue line corresponds to the arctangent ft function and the clear green line is the lorenztian ft function (Monteseguro et al., 2019).

characteristics. To predict the specific characteristics of a nanomaterial, the specific surface area (SSA) measurement or the total surface area per unit of mass is essential (Dollimore, Spooner, & Turner, 1976). When measuring surface area, gas adsorption techniques are typically used to determine the accessible surface. Methods can be used to determine the properties of the nanomaterials by gathering data on physio-sorption isotherms, which are relationships between the equilibrium pressure and volume of gas adsorbed. The Brunauer-Emmett-Teller (BET) method is the approach used most frequently to derive an SSA. The monolayer capacity of the gas on the sample is determined from the isotherm using a limited range of relative pressures. This information is then utilized to construct a BET specific surface area. In gas adsorption tests, the sample preparation is crucial to get rid of any adsorbed gases, water vapor or other unwanted surface impurities that might impair the gas adsorption. Furthermore, the BET method is a streamlined technique that makes more information about nanomaterial characterization which may not hold in all circumstances. As a result, additional verification may be required and approaches for boosting confidence can be explored elsewhere. With careful attention to sample preparation and the use of calibrants whenever it is practical, the BET method, despite its drawbacks, is a quick and affordable way to produce a relative assessment of the SSA of a wide variety of nanomaterials.

The fact that this measurement method can only be applied to dried samples is by far its biggest drawback. Because aggregation or agglomeration of particles may result in lower SSA values for tightly bound nanomaterials, it is necessary to perform a procedure on post-dried particles that facilitates an accessible surface area prior to measurement.

Figure 5.4a and 5.4b shows selected adsorption-desorption isotherms of $N_2$ at 77 K for activated carbons elaborated from vetiver roots and bagasse precursors, respectively. They give illustrative examples for the shape and behavior of the $N_2$ adsorption isotherms for these activated carbons (Passe-Coutrin et al., 2008).

## 5.6 THERMAL GRAVIMETRIC ANALYSIS (TGA)

Thermal analysis is one of the most useful methods for collecting physical and chemical data. TGA is probably the most widely used thermal analysis technique. TGA is used in various applications to provide information about the combination of components in the sample. Combined with other thermal analysis techniques such as differential scanning calorimetry (DSC) and spectroscopic techniques such as FTIR and MS, TGA can become a more powerful analytical technique. In a controlled environment, TGA measures the absolute amount and rate of change of sample weight as a function of time or temperature. TGA can measure a wide range of properties such as the thermal stability, oxidative stability, various atmospheric effects, humidity, and the presence of volatile components, and in some cases multi-component compositions. TGA determines that how the different components of a material are connected differently. When TGA is used in combination with DSC or Differential Thermal Analysis (DTA), the analysis mode is called the Simultaneous DSC-TGA (or DTA) (SDT). SDT, unlike the traditional DSCs, measures not only the amount and rate of weight change, but also the heat flow of the sample. SDT measures the same properties as the TGA, but adds the heat of reaction, melting point, and boiling point to the list. Weight, rate of weight

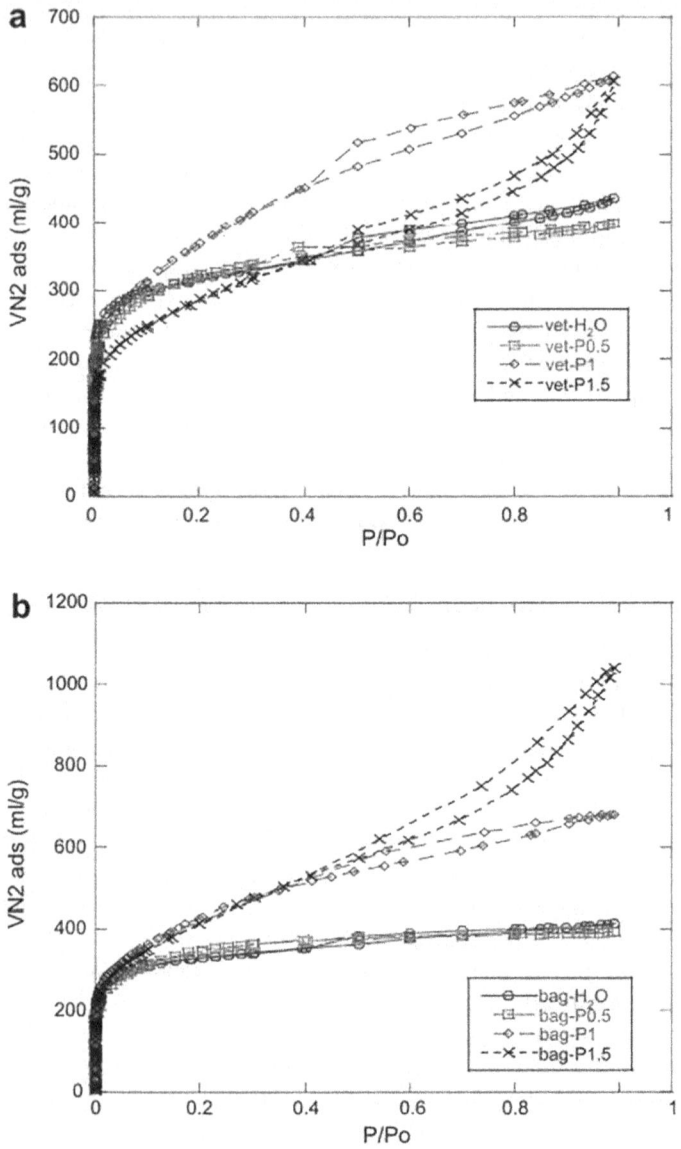

**Figure 5.4** Adsorption/desorption isotherms of $N_2$ at 77 K for steam and $H_3PO_4$ activated carbons derived from (a) vetiver roots and (b) bagasse (Passe-Coutrin, Altenor, Cossement, Jean-Marius, & Gaspard, 2008).

change-differential thermal analysis and temperature are the three most important signals captured by TGA when analyzing a sample. The first derivative of weight with respect to temperature or time shows a differential thermogravimetric (DTG) curve. DTG curves can provide both qualitative and quantitative data about the sample. Fingerprinting materials and distinguishing between two or more overlapping reactions is an example of a qualitative analysis method. The peak height and temperature of the peak weight loss measurement are examples of quantitative modes. Measurement accuracy is the most important aspect of MEP operation. Regular adjustments can help increase the reliability of the data collection. TGA requires both the mass and the temperature calibration. Most equipment and software packages include a semi-automatic mass calibration procedure in which the user places certified calibration weights on the equipment's sample platform. The Curie points of standard metal are used in the calculation of temperature calibration. The Curie point of a material is the temperature at which magnetic susceptibility is lost. To perform a Curie point temperature calibration, a strong magnet must be placed under or above the oven to cause an initial weight increase or decrease at room temperature (Gabbott, 2008).

Thermogravimetry (TG) and derivative thermogravimetry (DTG) were used to evaluate the weight loss and thermal characteristics of the white yam (*Dioscorea rotundata*) tuber peel (YTBS) residue adsorbent with the plots as shown in Figure 5.5. Thermogravimetric pyrolysis (under nitrogen) and combustion (in air) were programmed to work simultaneously at temperatures between 20°C and 500°C and 500°C and 600°C, respectively. Three sections (I, II, and III) on the TG curve (Figure 5.5a) represent the main stages in the degradation of the residue adsorbent. The DTG curve (Figure 5.5b) exhibits three exothermic peaks with maximum temperatures of 71°C, 281°C, and 509°C. The first region (I) relates to the removal of non-dissociative, physically absorbed, hydrogen bound water molecules from the residue surface and it occurred during the temperature ramp from 20°C to 100°C and the following hold at 100°C. By maintaining the temperature at 100°C for 60 minutes, total moisture evaporation in this area, which was responsible for 5% weight loss, was ensured. The volatile components of the lignocellulosic adsorbent are eliminated in the second stage (II), which has a weight loss of 66%. This can be attributed to the removal of the three main constituents; cellulose, hemicellulose, and lignin. These three components also have varying ranges of decomposition temperatures (hemicellulose, 150°C–350°C; cellulose, 275°C–380°C; and lignin, 300°C–500°C) and this can be related to their degree of thermal stability (Asuquo, Martin, & Nzerem, 2018).

At the end of this regime (II), the fixed carbon content of the YTBS adsorbent was 21%, which shows its intrinsic carbon content in the chars. The third region (III) in the TG curve (Figure 5.1a) shows the carbonization of the resulting chars from the pyrolysis regime (20°C–500°C) to obtain the residual ash composition of the YTBS adsorbent. This region is characterized by a maximum in the DTG curve (Figure 5.5b) between 503°C–530°C which corresponds to the decomposition of the chars in air (Asuquo et al., 2018).

## 5.7 SUPERCONDUCTING QUANTUM INTERFERENCE DEVICE MAGNETOMETRY (SQUID)

Measurements of magnetic fields and properties can be made using a variety of methods. Induction coils, flux gate magnetometers, reluctance and Hall Effect magnetometers, magnetic optical magnetometers, and optically pumped magnetometers have all been used as the basis of detection technology. Sensitivity levels range from Micro Tesla to Pico Tesla. The superconducting quantum interference device or SQUID is the most sensitive flux detector available. The device exhibited a field resolution of 1017 T while operating at very low temperatures with quantum limited sensitivity. Magnetization residue (MR), magnetization saturation (MS), and blocking temperature (TB) are typical SQUID measurements. SQUID can measure the magnetic response of a single molecule in addition to NPs. In fact, a scanning magnetic microscope with nanoSQUID mounted on sharp quartz has recently been developed. A very promising probe for magnetic spectroscopy and imaging at the nanoscale is NanoSQUID (Pappas, Prinz, & Ketchen, 1994). The nanoSQUID sensor is a deep submicron Josephson tunnel junctions provided by two Dayem nanobridges (superconducting membrane nanoshrinkage), which are made using focused ion beam (FIB) or electron beam lithography and have length and width that are similar to coherence lengths. An extremely small SQUID region is the primary need of SQUID intended for the detection of magnetic NPs. The loop size should preferably be comparable to the size of the NP immediately connected to it in order to achieve the best coupling coefficient. NanoSQUID offers the advantages of direct magnetization change measurements in small spin systems for magnetic

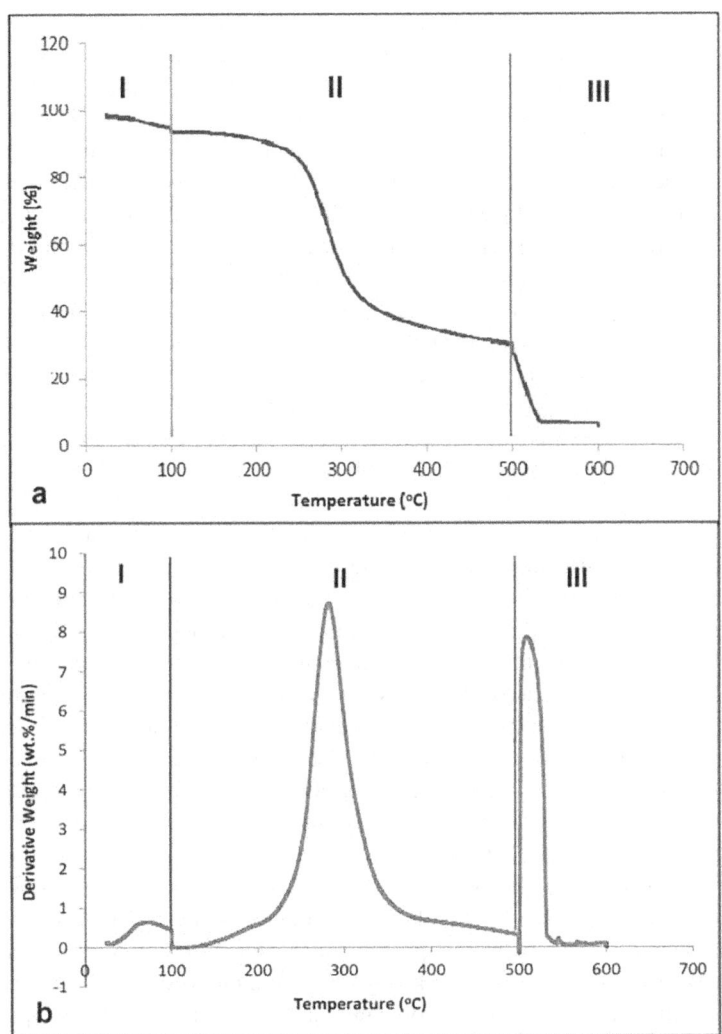

**Figure 5.5** Thermogravimetric (a) and derivative thermogravimetric (b) analysis of YTBS in $N_2$ (I-II) and air (III) environments (Asuquo et al., 2018).

resonance force microscopy or magneto-optical spin detection. The Dayem nanobridge from nanoSQUID is not only simple to create in a single nanopatterning step, but it also resists magnetic fields that are applied to the plane of the SQUID loop.

SQUID can only detect magnetic flux which is given by the product of BdA.(Kraft, Rupprecht, & Yam, 2017) This is equal to the magnetic flux density (B) multiplied by the area (A) of the SQUID loop. Since the effective area A of each SQUID is unknown, each SQUID magnetometer must be calibrated with a sample of known magnetic moments and therefore with a known fringe (or stray) magnetic field B. This is usually done by the manufacturer. Commercially available SQUID magnetometers typically detect changes in magnetic flux by mechanically moving the sample through a superconducting pickup coil. The pickup coil is converted to voltage VSQUID. The position is shown by Magnetic Properties Measurement System (MPMS) in the x direction parallel to the external magnetic field, $B_{ext}$, and by the raw data which produces the so-called "last scan." The pickup coil is designed as a secondary gradient meter to ensure that the effects of all types of external magnetic fields are suppressed. A single SQUID scan with the maximum VSQUID on x-pos while approximately 2 cm corresponds to a sample placed directly between the dual coils of the pickup gladiolus. This center position must be determined by a long scan after installing a new magnetic sample in order to correctly set the sample position with respect to the pickup coil. It has to be noted that in actual measurements it is usually best to perform a relatively long scan so that the scan includes not only the maximum value but also the two minimum values. In fact, if the sample is an ideal point dipole that is precisely placed on the axis of the magnetometer, a single SQUID scan will be fully adjusted automatically without the user having direct access to the routine. With standard software, two types of fitting are possible. The fitting curve can start from a fixed sample position or fit the amplitude of the VSQUID (x-pos.) curve with a single fitting parameter, which is the magnetic moment of the expected point-of-sale dipole. Alternatively, use MPMS. This is known as a linear regression mode. Alternatively, in iterative regression mode, you can adjust the position of the sample along with the amplitude. When recording the temperature dependence, iterative regression mode easily compensates for thermal expansion of the sample holder assembly. The sample size should be limited to 5 mm or less along the scan direction to reduce the fitting errors caused by the point dipole approximation in both fitting routines. The sample holder, which is usually a transparent straw, limits the sample size to 5–6 mm. However, it has long been observed that the point dipole approximation does not have the correct magnetic moment value for the system (Clarke & Braginski, 2004).

## REFERENCES

Asuquo, E. D., Martin, A. D., & Nzerem, P. (2018). Evaluation of Cd (II) ion removal from aqueous solution by a low-cost adsorbent prepared from white yam (Dioscorea rotundata) waste using batch sorption. *ChemEngineering, 2*(3), 35.

Bugnet, M., Overbury, S. H., Wu, Z. L., & Epicier, T. (2017). Direct visualization and control of atomic mobility at {100} surfaces of ceria in the environmental transmission electron microscope. *Nano Letters, 17*(12), 7652–7658.

Clarke, J., & Braginski, A. I. (2004). *The SQUID handbook* (Vol. 1): Wiley Online Library.

Dollimore, D., Spooner, P., & Turner, A. (1976). The BET method of analysis of gas adsorption data and its relevance to the calculation of surface areas. *Surface Technology, 4*(2), 121–160.

Fendler, J. H. (2008). *Nanoparticles and nanostructured films: preparation, characterization, and applications*: John Wiley & Sons.

Filipe, V., Hawe, A., & Jiskoot, W. (2010). Critical evaluation of nanoparticle tracking analysis (NTA) by NanoSight for the measurement of nanoparticles and protein aggregates. *Pharmaceutical Research, 27*(5), 796–810.

Gabbott, P. (2008). *Principles and applications of thermal analysis*: John Wiley & Sons.

Guehrs, E., Schneider, M., Günther, C. M., Hessing, P., Heitz, K., Wittke, D., Eisebitt, S. (2017). Quantification of silver nanoparticle uptake and distribution within individual human macrophages by FIB/SEM slice and view. *Journal of Nanobiotechnology, 15*(1), 1–11.

Joosten, J. G., McCarthy, J. L., & Pusey, P. N. (1991). Dynamic and static light scattering by aqueous polyacrylamide gels. *Macromolecules, 24*(25), 6690–6699.

Kalinin, S. V., & Balke, N. (2010). Local electrochemical functionality in energy storage materials and devices by scanning probe microscopies: status and perspectives. *Advanced Materials, 22*(35), E193-E209.

Kraft, A., Rupprecht, C., & Yam, Y.-C. (2017). Superconducting quantum interference device (SQUID). *UBC Phys.*

Lendaris, G. G., & Stanley, G. L. (1970). Diffraction-pattern sampling for automatic pattern recognition. *Proceedings of the IEEE, 58*(2), 198–216.

Li, T., Senesi, A. J., & Lee, B. (2016). Small angle X-ray scattering for nanoparticle research. *Chemical Reviews, 116*(18), 11128–11180.

Minelli, C., Sikora, A., Garcia-Diez, R., Sparnacci, K., Gollwitzer, C., Krumrey, M., & Shard, A. G. (2018). Measuring the size and density of nanoparticles by centrifugal sedimentation and flotation. *Analytical Methods, 10*(15), 1725–1732.

Monteseguro, V., Sans, J., Cuartero, V., Cova, F., Abrikosov, I. A., Olovsson, W., Jönsson, H. J. M. (2019). Phase stability and electronic structure of iridium metal at the megabar range. *Scientific Reports, 9*(1), 1–9.

Nishiyama, H., Suga, M., Ogura, T., Maruyama, Y., Koizumi, M., Mio, K., Sato, C. (2010). Reprint of: Atmospheric scanning electron microscope observes cells and tissues in open medium through silicon nitride film. *Journal of Structural Biology, 172*(2), 191–202.

Pappas, D., Prinz, G., & Ketchen, M. (1994). Superconducting quantum interference device magnetometry during ultrahigh vacuum growth. *Applied Physics Letters, 65*(26), 3401–3403.

Passe-Coutrin, N., Altenor, S., Cossement, D., Jean-Marius, C., & Gaspard, S. (2008). Comparison of parameters calculated from the BET and Freundlich isotherms obtained by nitrogen adsorption on activated carbons: a new method for calculating the specific surface area. *Microporous and Mesoporous Materials, 111*(1–3), 517–522.

Reimer, L. (2013). *Transmission electron microscopy: physics of image formation and microanalysis* (Vol. 36): Springer.

Thomas, J. M., Midgley, P. A., Ducati, C., & Leary, R. K. (2013). Nanoscale electron tomography and atomic scale high-resolution electron microscopy of nanoparticles and nanoclusters: a short surveyNanoscale electron tomography and atomic scale high-resolution electron microscopy of nanoparticles and nanoclusters: a short surveyretain. *Progress in Natural Science: Materials International, 23*(3), 222–234.

Verbeni, R., Sette, F., Krisch, M., Bergmann, U., Gorges, B., Halcoussis, C., Ruocco, G. (1996). X-ray monochromator with 2 × 108 energy resolution. *Journal of Synchrotron Radiation, 3*(2), 62–64.

Zhang, Z. M., Zhang, Z. M., & Luby. (2007). *Nano/microscale heat transfer* (Vol. 410): Springer.

Zizzari, A., Bianco, M., Carbone, L., Perrone, E., Amato, F., Maruccio, G., Arima, V. (2017). Continuous-flow production of injectable liposomes via a microfluidic approach. *Materials, 10*(12), 1411.

# INDEX